Radio Spectrum Conservation

Radio Spectrum Conservation

William Gosling

OXFORD AUCKLAND BOSTON JOHANNESBURG MELBOURNE NEW DELHI

Newnes
An imprint of Butterworth-Heinemann
Linacre House, Jordan Hill, Oxford OX2 8DP
225 Wildwood Avenue, Woburn, MA 01801-2041
A division of Reed Educational and Professional Publishing Ltd

 A member of the Reed Elsevier plc group

First published 2000

British Library Cataloguing in Publication Data
A catalogue record for this book is available from the British Library

ISBN 0 7506 3740 4

Library of Congress Cataloguing in Publication Data
A catalogue record for this book is available from the Library of Congress

Typeset by David Gregson Associates, Beccles, Suffolk
Printed in Great Britain by Biddles Ltd, Surrey

Contents

PREFACE

In the early 1970s the Club of Rome published its famous document on *Limits to Growth*, which argued that the expansion of world economic activity could not continue indefinitely because many natural resources would rapidly be exhausted. Although many of the specific predictions made at that time have not come to pass, nevertheless this publication marked the beginning of a general concern in the developed world about the prospect that scarce and non-renewable resources, such as oil and vital minerals, might be moving to exhaustion, even if at a more remote date than was at first thought.

Curiously enough, in the wide-ranging debate that followed the Club of Rome's bombshell scarcely any attention was paid to the resource which is (of all natural resources) the most certainly limited and the least susceptible of economic substitution – that part of the electromagnetic spectrum suited to radio transmission. The radio spectrum is fixed in extent by unalterable physical realities, and neither optics nor cables can substitute for radio in most of its important applications. The increasing use of radio correlates very closely with world economic development, and congestion of the radio spectrum in many bands is now a fact of life in all populated areas of developed countries.

This places a very heavy responsibility on radio engineers, who must design new radio systems which satisfy the ever-growing demands that can be accommodated within the strict limits of the available spectrum. It is also necessary to wage a never-ending battle against radio spectrum pollution, which can destroy a precious resource to no positive end whatsoever. The ethical dimension of this problem is clear: profligate use or pollution of the electromagnetic spectrum potentially harms the whole of human kind. We can no longer plead ignorance of the problem; it is now universally recognized. Even in those parts of the world where the spectrum is, as yet, uncluttered we must not repeat the mistakes of the past, which either lead to a closing-off of opportunities for the future or to massive waste of investment, and sometimes both. To be specific, the days when we could be easy in our consciences about radiating powerful analogue FM signals from high masts are long gone, and that is but the most obvious example.

In future, socially responsible radio engineers must ask themselves not only whether proposed future developments are technically feasible and economically viable but also whether they are morally acceptable. This book tries, building from first principles, to explain what the basics of ethical radio design are.

It has not been an easy book to write because it looks at many problems from a radically different standpoint, and I owe a debt of thanks to Matthew Deans for his encouragement during its long gestation. I also acknowledge my gratitude to Richard Hillum, John Gardiner, Joe McGeehan, Bruce Lusignan, John Martin and the late Vladimir Petrovic for helpful discussions.

William Gosling

Part One

The Radio Spectrum as a Finite Resource

CHAPTER 1

THE SIGNIFICANCE OF RADIO

In 1887 a radio transmitter began working briefly, the first and at that time the only one anywhere in the world. It was located in the research laboratory of a young German physicist, Heinrich Hertz, and its transmissions could be received over a range of just a few metres. Today, by contrast, there are literally tens of millions of radio transmitters worldwide, some of them having a short range like his, but others with global and even extraterrestrial reach.

1 Evolution of radio technology

The vigorous growth in radio use continues. New services often expand spectacularly. Soon after the introduction of cellular radiotelephones in Europe growth rates as high as 40% per year were exceeded in several countries. At this rate of expansion the number of users doubles about every year and a half. There is a smaller but still substantial rate of growth in all other personal and mobile radio use, with doubling times around four years. Air traffic is growing rapidly worldwide, and in consequence the need for new communication channels to aircraft and for additional radar increases at least as fast. Passenger telephone and data connections are becoming increasingly important in all kinds of transportation, land, sea and air. At the same time the military require increasingly sophisticated and extensive radio systems, as also do the police, fire and ambulance services. Now not merely voice circuits but also picture and data transmission are routine in all the emergency services, along with telemetry and remote control.

As for broadcasting, the television service has been launched on a precipitate development, thanks to satellite, cable and above all digital television, and the number of channels available to the typical user is set to multiply one hundred-fold between 1990 and 2010. Even the oldest established services, such as sound broadcasting in the medium wave (MF) band, show every sign of pent-up demand. Whenever the opportunity to set up a new station arises many contractors compete for the licence.

In short, the utilization of radio increases as an inevitable consequence of social and economic development, and rather faster than most other growth indices. Yet all of this activity has to be contained within a small part of the electromagnetic spectrum, the **radio bands**, and these can never be extended for reasons which will be made clear later. There is therefore a classic conflict between the needs of a growing and developing human society and a finite resource – the electromagnetic spectrum – with which those needs must be met. Indeed there is a real danger that the radio spectrum will be the first of our finite resources to run out, long before oil, gas or mineral deposits.

How, then, can we supply the radio-based services which our society demands within these limitations? Beyond question this is the most difficult challenge facing radio engineers at the present time, and this book describes the various ways in which it is tackled and can be resolved. The obvious starting point is to look at the resource which is available, and it is convenient to do so by looking at how we got to where we are today.

1.2 What is the radio spectrum?

Before we can speak about the radio spectrum it seems reasonable to ask what radio energy itself is. Historically, there was a great debate between those who believed, like Isaac Newton (1642–1727), that light energy (and therefore later radio also) was a stream of **particles**, and those, following Christiaan Huygens (1629–95), who thought that it was **waves**. Now we know that it is composed of particles (**photons** or **quanta**) which have wave properties, so there was something to be said for both opinions.

In any event, we certainly take the view that it is electromagnetic. The first person to observe a connection between electricity and magnetism was Hans Christian Öersted (1777–1851) who in 1820 found that a magnetized compass needle moved when an electric current flowed in a wire close to it. The effect was studied experimentally by André Marie Ampère (1775–1836) in France, Joseph Henry (1797–1878) in the United States and Michael Faraday (1791–1867) in England. Faraday obtained detailed experimental evidence for the ways in which magnetic fields and electric currents could interact; however, fully developing the theory proved beyond him, and the task was taken up by James Clerk Maxwell (1831–79), a Scot and one of the greatest theoretical physicists of all time.

Being uncomfortable about the notion of forces somehow acting on things situated at a distance from each other, with nothing in between to communicate them, Maxwell chose to look at electromagnetic phenomena as manifestations of stresses and strains in a continuous elastic medium (later called the electromagnetic **aether**) that we are quite unaware of, yet which fills all the space in the universe. Using this idea, Maxwell was able to develop an essentially mechanical model of all the effects Faraday had observed so carefully. Between 1864 and 1873 Maxwell demonstrated that a few relatively simple equations could fully describe electric and magnetic fields and their interaction. These famous equations first appeared in his book *Electricity and Magnetism* published in 1873. Developed empirically, Maxwell's theory predicted accurately all the electromagnetic effects that could be observed in his time, and of course it still works very well in the majority of situations, although as we now know it will fail where quantum effects become significant, for example in photoelectricity.

Maxwell used his equations to describe the electromagnetic field, how it is produced by charges and currents, and how it is propagated in space and time. The electromagnetic field is described by two quantities, the electric component **E** and the magnetic flux

B, both of which change in space and time. The equations (in modern vector notation) are:

$$\nabla \cdot \boldsymbol{D} = \rho \qquad \rho \text{ is electric charge density} \tag{1.1}$$

$$\nabla \cdot \boldsymbol{B} = 0 \tag{1.2}$$

$$\nabla \times \boldsymbol{E} = -\partial \boldsymbol{B}/\partial t \tag{1.3}$$

$$\nabla \times \boldsymbol{H} = \boldsymbol{J} + \partial \boldsymbol{D}/\partial t \qquad \boldsymbol{J} \text{ is electric current density} \tag{1.4}$$

The electromagnetic effects observed experimentally by Faraday (and many more beside, but not quite all) can be predicted theoretically by these equations, a great triumph for Maxwell. He also calculated that the speed of propagation of an electromagnetic field is the speed of light, always represented as c, and concluded that light is therefore an electromagnetic phenomenon, although visible light forms only a small part of the entire spectrum.

This analysis also leads to the most fundamental of all equations in radio

$$c = f \cdot \lambda \tag{1.5}$$

where f is frequency, and λ is wavelength.

Maxwell's conclusions, that light consists of electromagnetic waves, were in line with the scientific beliefs of his time, and seemed confirmed experimentally by (among other things) the fact that the wavelength of light had been successfully measured many years before. It had been found as early as the 1820s that violet light corresponded to a wavelength of about 0.4 microns, orange–yellow to 0.6 microns and red to 0.8 microns, all of which fitted perfectly with Maxwell's ideas.

It was an obvious further consequence of his theory that there might also be waves of much greater length (and correspondingly lower frequencies). Maxwell confidently predicted their existence, even though they had never been observed. He died (1879) before there was experimental confirmation of this radical insight. In 1887 Heinrich Hertz (1857–94) was the first to demonstrate the existence of radio energy experimentally. As his transmitter he used a spark gap connected to a resonating circuit, which determined the frequency of the electromagnetic waves and also acted as the antenna. The receiver was a very small spark gap, also connected to a resonant circuit and observed through a microscope, so that tiny sparks could be seen. Hertz generated radio energy of a few centimetres wavelength and was able to demonstrate that the new waves had all the characteristics previously associated exclusively with light, including reflection, diffraction, refraction and interference. He also showed that radio waves travel at the speed of light, just as Maxwell had predicted.

After Maxwell's early death, Albert Michelson and Edward Morley devised experiments (1881, 1887) which showed that the aether he had assumed does not exist, thus demolishing the physical basis of his theories. However, although the physical ideas Maxwell used were wrong, the equations were empirically based and necessarily remained an excellent fit to observations (in all but a very few cases). They continued to give the right answers, even though the path to them was discredited, and they remain very widely used to this day.

1.3 A quantum interpretation

We now know that there is no elastic medium for the waves to propagate in, so it follows that the waves Hertz thought he had discovered are not at all what he supposed them to be. A wholly new theory is needed. The modern answer is that radio energy, in this respect closely similar to light, propagates freely in space as a stream of very small, light particles called electromagnetic quanta or photons (Gosling 1998). The difference between the quanta of light and radio is solely that each quantum of light carries far more energy than those of radio, but in other ways they are identical.

However, because these particles are very small indeed they do not obey classical mechanics (Newton's laws), as do snooker balls, for example. Instead they behave in accordance with the laws of quantum mechanics, like all very small things. This gives them some strange properties, quite unfamiliar in everyday life, which may seem contrary to common sense. Two are important:

1. The first is that radio quanta can exist only when in motion at the velocity of light. In free space this velocity is 299 792 456.2 metres per second, but 300 million (3×10^8) metres per second is a very good approximation for all but exacting situations.

2. The second of these strange properties is that particles as small as radio quanta also have wave-like properties. All things which are small enough have noticeable wave-like properties, even particles of matter. However, as things get bigger their wave properties get less perceptible, which is why we do not notice them in ordinary life.

We can characterize the state of a particle if we specify its energy or momentum, while for a wave the corresponding parameters are frequency and wavelength. Quantum mechanics relates these pairs of parameters together, linking the wave and particle properties of quanta, in two monumentally important equations:

$$\mathcal{E} = hf \qquad (1.6)$$

where h is Planck's constant, and $\mathcal{E}$ is energy.

$$m = h/\lambda \qquad (1.7)$$

where m is momentum.

Planck's constant, relating the wave and particle sides of the quanta, is one of the constants of nature, and has the amazingly small value 6.626×10^{-34} J/s. The tiny magnitude of this number explains why the classical theories work so well. Quanta have such very small energy (and hence mass, since $\mathcal{E} = mc^2$) and in any realistic rate of transfer of energy (power flow) they are so very numerous that their individual effects are lost in the crowd and all we see is a statistically smooth average, well represented by the classical theory.

In quantum mechanics the correspondence principle states that older valid classical results remain valid under quantum mechanical analysis. (But the latter can also reveal things beyond the classical theory.) Maxwell's equations work as well as ever they did, but it is good to know what is really going on (which is quite different from what Maxwell imagined) and there are times when thinking about what is happening to the quanta can actually help a better understanding.

When there is a flow of quanta all of the same frequency (and hence energy), the radiation is referred to as **monochromatic** (if it were visible light it would all be of one colour), and if it all comes from a single source, so that the quanta all start out with their wave functions in phase (or at least phase-locked), the radiation is said to be **coherent**. Radio antennas produce coherent radiation, as (at a very different wavelengths) do lasers, but hot bodies produce incoherent radiation, experienced in radio systems as noise. By contrast, incoherent radiation is fascinating to radio astronomers, for whom hot bodies are primary sources.

In the case of coherent radiation very large numbers of radio quanta are present, but the wave functions associated with each photon (quantum) have the same frequency and are in a fixed phase relationship, so we can treat them as simply a single electromagnetic wave, which is why Maxwell's mathematical theory works so well in practice.

1.4 The electromagnetic spectrum

Hertz confirmed Maxwell s prediction that electromagnetic energy existed not only as light but also in another form with much longer wavelengths (what we would now call radio). As a result the idea of an electromagnetic spectrum quickly developed.

For centuries people had known that the sequence of colours in the light spectrum was red, orange, yellow, green, blue then violet. By the nineteenth century this had been associated with a sequence of reducing wavelengths (or increasing frequencies) from the long-wave red to the short-wave violet. Invisible infrared waves, longer in wavelength than red, had been discovered, as also had the

ℰ 10^-33 10^-30 10^-27 10^-24 10^-21 10^-18 10^-15 J
f 10^0 10^3 10^6 10^9 10^12 10^15 10^18 Hz
the radio bands
IR
UV
X rays
λ 10^8 10^5 10^2 10^-1 10^-4 10^-7 10^-10 m
IR = infrared
UV = ultraviolet
= visible light
known in
Maxwell's day

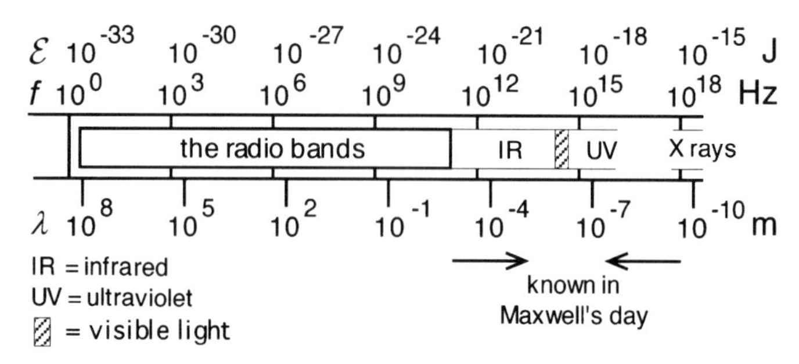

Fig. 1.1
The electromagnetic spectrum.

ultraviolet, shorter than violet. Now it was possible to imagine that electromagnetic waves might extend to much longer wavelengths than infrared. Later, when X-rays were identified, it was also possible to fit them in as electromagnetic waves even shorter than ultraviolet.

Thus it became possible to see all the forms of electromagnetic energy as a continuous spectrum (Fig. 1.1). Quantum mechanics has not overturned this picture, but at each frequency we now add a particular value of quantum energy.

Radio technology is concerned with the lower (in frequency) part of the electromagnetic spectrum. As a matter of convenience, the radio part of the electromagnetic spectrum is further subdivided into a series of bands, each covering a 10 : 1 frequency ratio, as in Fig. 1.2. Except in the uppermost (EHF) band, the quanta are insufficiently

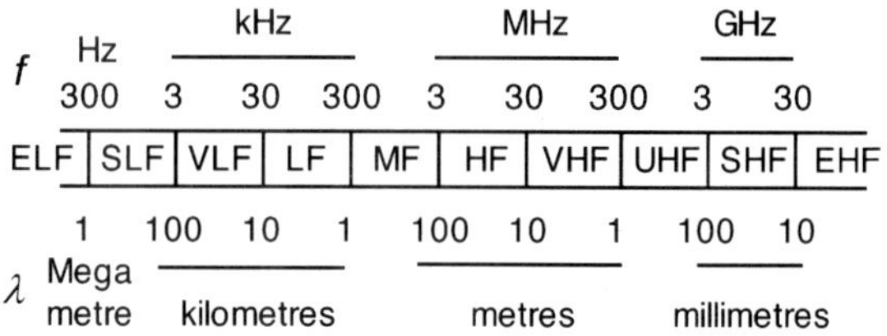

kHz
MHz
GHz
f
Hz
300 3 30 300 3 30 300 3 30
ELF SLF VLF LF MF HF VHF UHF SHF EHF
1 100 10 1 100 10 1 100 10
λ
Mega
metre
kilometres
metres
millimetres

Fig. 1.2
The radio bands

energetic to interact with the gas and water vapour molecules of the atmosphere, which is therefore transparent to radio signals. This is a great practical advantage and allows radio signals to propagate over long distances.

1.5 Limits on the radio spectrum

The radio bands extend from frequencies of 30 Hz (ELF) to 300 GHz (EHF). Why are there no radio bands outside this frequency range? The reasons stem from basic physical considerations, and will never be overcome.

The lower limit is already not very far from zero frequency and is set by antennas and available bandwidth. The efficiency of an antenna in radiating radio energy is dependent on its length expressed as a fraction of a wavelength, so at the lowest frequencies antennas have to be made very large if they are to achieve even very low efficiency. For example, the 'Sanguine' submarine communications system (operating in the ELF band at about 30 Hz or 10 000 km wavelength) was designed (but never built) for the US Navy. It was to have had as a base antenna a 25 km square mesh of wires. With an RF input of 10 MW, calculations showed that it would have radiated only 147 watts, all the remainder of the power being dissipated as heat. Evidently the use of bands even this low is all but impracticable.

But there is worse: to extend the radio bands lower in frequency would be pointless because at the lower end of the radio spectrum there is so little information-carrying capacity. The whole VLF band, for example, is only 27 kHz wide, scarcely more than a single UHF voice channel in many systems, and lower bands are even more constricted. Furthermore, any resonant system or circuit in transmitter or receiver (such as the antenna, which must be tuned to resonate to give best efficiency) results in very narrow bandwidth, even if Q-factor is low. So in these bands the data rates of transmissions are unavoidably very low; indeed ELF transmissions to submerged submarines are said to operate at only one bit per second. In this particular case it may not matter, because the messages transmitted are doubtless brief, but the general utility of systems of this kind would be vanishingly small. It is clear, then, that there is not the slightest prospect of extension of the radio bands downward.

The upper frequency limit on radio use is set mostly by the increasing opacity of the atmosphere to radio transmissions. This is due to interactions between radio transmissions and molecules of water or oxygen, which absorb the radio quanta. Resonance effects occurring in molecules and atoms can greatly increase their apparent cross-section for capturing quanta near their various resonance frequencies. The dimensions of the molecules are such that these resonances become significant at frequencies above a few tens of gigahertz for the normal atmospheric gases. In consequence above 30 GHz atmospheric absorption increases rapidly and steadily although with localized maxima at particular frequencies, for example reaching 14 dB/km in the first oxygen absorption band at 60 GHz (Fig. 1.3). In addition to local maxima, however, the basic

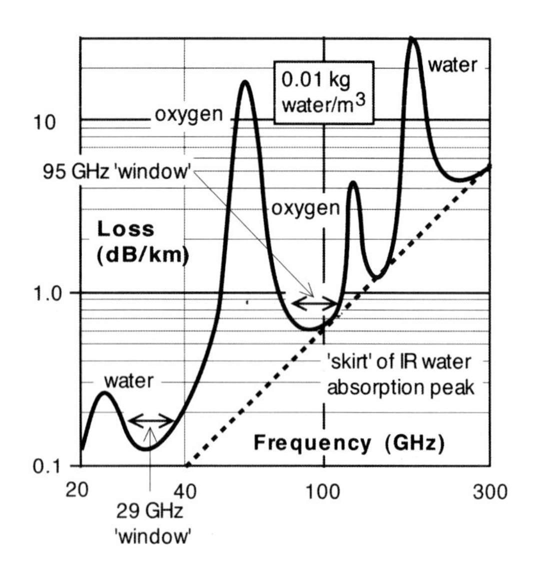

Fig. 1.3
Absorption by the atmosphere. (Source: van Vleck, J.H. *Phys. Rev.* **71** (1947))

curve of absorption also rises steadily, with the result that above 300 GHz the radio quanta are unable to travel more than a few metres in air before being absorbed, and radio technology becomes ineffective. Even below this absolute cut-off frequency the shorter wavelengths (less than 10 mm, corresponding to frequencies above 30 GHz) present increasing problems to users.

Above all, the problem of absorption of the quanta by water vapour makes communication very dependent on weather conditions. Other difficulties arise because of the quasi-optical propagation near the upper frequency limit. Diffraction of quanta around radio-opaque objects (which means most things in the natural and human-built environment near the Earth's surface) extends for a distance into the 'shadow' behind them which is directly dependent on the wavelength of the quanta. Thus for short wavelengths in typical human environments deep radio shadows behind obscuring objects create problems for reliable coverage. Worse still, most building and structural materials have increasing absorption into the EHF, so there is poor building and vehicle penetration at these frequencies. Thus there is only a limited prospect for use of the EHF bands, and from present perspectives none at all for radio techniques above 300 GHz. No higher radio bands are in use, and no other 'window' of transparency appears in the atmosphere until some hundreds of times higher in frequency, corresponding to the infrared, visible and ultraviolet light regions.

These factors together set the natural upper and lower boundaries on the usable radio bands, which justifies the treatment of the radio spectrum as a strictly limited, and hence increasingly valuable, natural resource. There will never be any more of it. In practice, under the ordinary economic and social constraints within which we must all live, the use of radio bands proves even more severely restricted than this discussion indicates. At present only specialized services operate outside 100 kHz to 35 GHz and a substantial part of the more socially important applications (other than radar and satellites) are crowded into the VHF and UHF bands, which have many particularly favourable properties from the user's standpoint. Continuing world economic development, of which radio use is an integral and inescapable part, therefore depends on deploying this increasingly strained resource in a way that uses technical ingenuity to avoid its effective exhaustion. It seems likely that the radio spectrum will potentially be the first of the finite resources to limit the development of our civilization, long before energy or materials exhaustion become significant, so spectrum conservation is the most pressing task facing radio engineers, not only now but for the foreseeable future.

1.6 Problems of spectrum use

The rapidly growing congestion of the bands leads inexorably to intensifying problems. With so many radio stations in competition for a limited resource, how is it possible to receive just the particular transmission that the user has chosen, yet nothing else at the same time which might impair the wanted signals? The difficulties arising have two critical aspects: **electromagnetic pollution** and **interference**.

Electromagnetic pollution consists of all radio energy radiated other than as a necessary part of a legitimate transmission. Some of this is natural and unavoidable, like the radio quanta emitted during lightning strikes, heard as crashes and bangs on LF and MF broadcast receivers. However, the most troublesome forms of electromagnetic pollution are made by human agency. Much of the electrical equipment in our homes and places of work emits spurious radio energy unless precautions are taken to stop it doing so. Electric motors, relays and contactors, above all the spark-ignition systems of internal combustion engines, all these are copious sources of radio energy in many bands unless suitable steps are taken. A generation ago this presented a serious problem to the radio engineer, but today in all developed countries there are strict regulations calling for the incorporation of radio suppression components in all such equipment.

Paradoxically, the principal residual source of human-made pollution is radio transmitters and their associated antennas, which in addition to radiating their legitimate signals also radiate other radio energy, the **spurious outputs**, which act simply as an electromagnetic pollutant. For example, transmitters may radiate significantly at frequencies which are harmonics of those for which they are designed. They may also have spurious emissions at other, unrelated frequencies and in the frequency channels adjacent to those in which they operate. All of these effects are minimized by careful equipment design – 'good housekeeping' with all radio, electrical and electronic equipment. In principle spectrum pollution ought not happen at all, and in practice it is indeed possible to keep it to an acceptably low level, with care.

Interference, by contrast, occurs when energy from two or more legitimate transmissions reaches the receiver, but only one is wanted. The transmissions may activate the main response of the receiver, or may affect its spurious responses, which cause it to give an output in certain circumstances despite the input not being in the chosen form. Radio interference is potential in all use of radio communications, and its management presents a much more difficult problem than pollution. The solution of the interference problem is key to the continued successful exploitation of radio technology.

These important problems of radio use will need to be considered individually. In order to do so, however, it will first be necessary to summarize the phenomena of radio propagation (Gosling 1998).

Questions

1. A space probe weighing 10 kg is stationary in space relative to a certain reference point. It transmits toward that point at a power level of 1 kW for 1000 seconds. How far will it move and in what direction in the next 1000 seconds? (0.33 m away from the reference point)

2. Why is it impossible to increase the amount of radio spectrum available for use?

3. How would the usable radio spectrum change in the case of operations in space, remote from planetary surfaces?

4. Why are car engines required by law to be fitted with ignition interference suppression?

Chapter 2

Radio propagation

Radio quanta, like all photons, have no electric charge and very little mass, so they move in straight lines in all but the very strongest gravitational fields. In the usual case quanta may be taken as originating from point sources, since the transmitting antennas are generally far away from the point at which the quanta are received and therefore of negligible dimensions.

2.1 The isotropic radiator

In the idealized theoretical case antennas may be considered which are **isotropic**, that is they radiate photons uniformly in all directions. This leads to a very simple theoretical treatment. Real world antennas are not like this, however, but are non-uniform radiators, having a **polar diagram** which defines a more or less complex variation of the density of emission of radio quanta with angle, ranging from the 'doughnut' shape of the simple dipole to the pencil beam of a paraboloid reflector (Gosling 1998). Even so, it would be desirable in some way to represent a real antenna by an isotropic equivalent, so that the simplest possible theoretical treatment of radio propagation could still be used. The first step is therefore to reconcile the complexities of actual antennas with this simple theoretical model.

In any particular direction the density of radio quanta radiated by a real antenna is exactly the same as it would be if an **isotrope** were

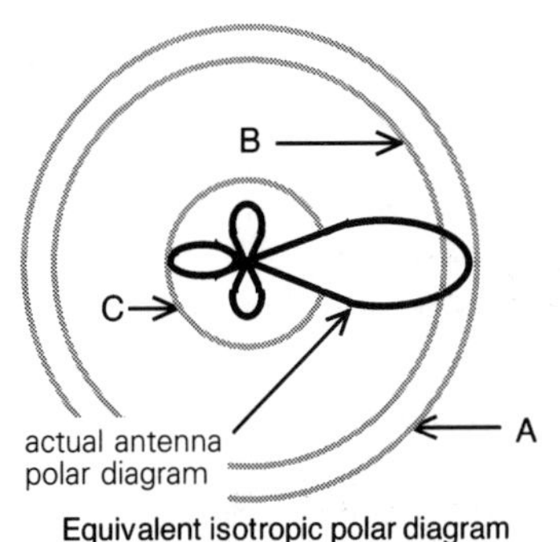

Fig. 2.1
Equating an actual polar diagram to its isotropic equivalent.

used with its power equal to the actual total power radiated by the real antenna multiplied by the gain of the antenna in the direction concerned (Fig. 2.1). In the direction of the peak of the main lobe of the antenna polar diagram an isotropic radiator with polar diagram A would produce an identical flow of radio quanta (but only in that one direction) to the antenna actually used. Radiation in the directions of the half-power points on the main lobe will correspond to that of the less powerful isotrope **B**, while in the reverse direction it would be the much less powerful isotropic radiator represented by C which equates to the actual antenna.

In any specified direction there is always an equivalent isotropic radiator which produces a radio quantum flow identical with that of the actual antenna. However, this equivalent isotropic radiator must radiate a different total number of radio quanta when integrated over the whole sphere, equating to a different total power. This notional power radiated by the isotrope which produces, in a particular direction, the same radio quantum flux as the actual antenna is called the **EIRP (equivalent isotropic radiated power)**. Obviously, if the EIRP is P_i while P is the total power radiated by the actual antenna, then if G is the power gain:

$$P_i(\phi, \theta) = G(\phi, \theta) \cdot P \tag{2.1}$$

In words, the EIRP in a certain direction is the actual power multiplied by the antenna power gain in that direction. The EIRP concept is useful because using it we can develop propagation theory assuming the simple case of isotropic radiators, and then get the right answers simply by using the EIRP as the isotrope power in place of the actual transmitter power.

Sometimes the EIRP is specified without giving the polar co-ordinates which indicate in which direction it is to be taken. In such cases it is the accepted convention that the direction intended is that of the peak of the main lobe. Sometimes this is called the **main lobe EIRP**.

2.2 Propagation in space

The simplest of all possible radio propagation environments is free space. Neither atmosphere nor any solid objects are present to complicate matters. Without loss of generality, because we have established the concept of EIRP, we can now develop a theory for this case assuming isotropic radiators.

Think of a small area ΔA distant r from an isotrope at point T, which emits a total power P_i (Fig. 2.2). The radio quanta are emitted uniformly in all directions, and therefore if the total

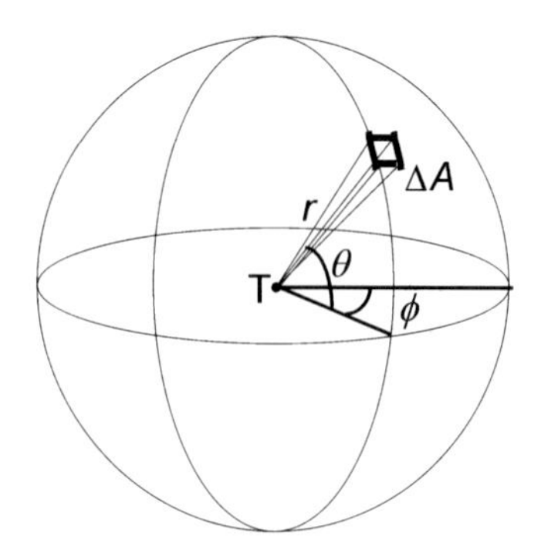

Fig. 2.2
Isotropic propagation in free space.

number of photons emitted per second is N, each having energy hf, the number passing through ΔA is equal to ΔN per second where:

$$\Delta N = \frac{N \cdot \Delta A}{4\pi r^2} = \frac{\Delta A}{4\pi r^2 hf} \cdot P_{\mathrm{i}} \tag{2.2}$$

Sometimes instead of the number of photons passing through this small area we may prefer to see the power flow. This is the number of radio quanta per second multiplied by the energy of each radio quantum, and is thus:

$$\Delta P = \frac{\Delta A}{4\pi r^2} \cdot P_{\mathrm{i}} \tag{2.3}$$

Note the **inverse square law** – just a consequence of the fact that in space the radio quanta do not get 'lost' by any mechanism and are therefore simply a constant number spread over a sphere which increases in area with the square of its radius. All propagation in space (or closely similar situations) is characterized by this inverse square law.

If an atmosphere is present between the transmitting and receiving antennas it is still usual to speak of space-wave propagation provided that the proximity of the Earth or other material objects does not affect the propagation. In this situation the inverse square law still holds.

2.3 The atmosphere

For radio transmissions within the atmosphere two additional important effects may occur, namely refraction and absorption. Refraction is bending of the path taken by radio photons, and occurs because the speed of a radio quantum which travels surrounded by matter, even a gas, is less than it would be in free space, due to a retarding effect by the surrounding atoms. The Earth's atmosphere has a density, and therefore a refractive index, largest near the surface and falling with height. It acts somewhat like a prism, tending to bend the path of radio quanta down toward the surface, so that the quanta follow the curvature of the Earth to

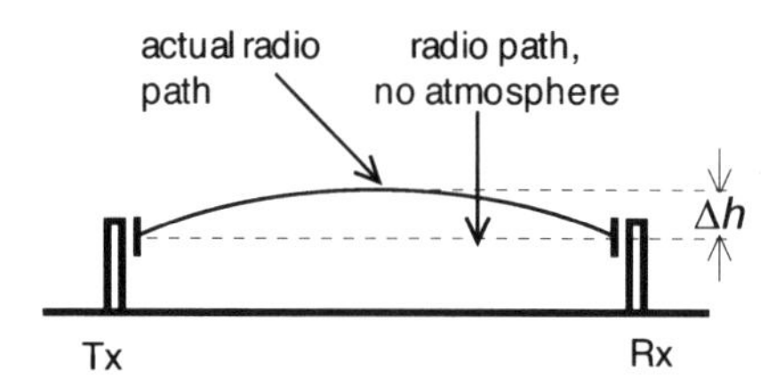

Fig. 2.3
The refracted radio path relative to the Earth's surface (curvature exaggerated for clarity).

some extent (Fig. 2.3). This has traditionally been analysed as equivalent to photons taking a straight path over an Earth of reduced curvature (increased radius) by a factor close to 4/3. It results in a modest but useful increase in the range of radio transmissions, particularly when a radio path grazes the top of a hill or ridge. The coverage area of broadcasting stations is similarly enhanced. However, this analysis assumes that the air is uniform in its properties, which thus vary significantly only with altitude, due to the change in pressure and temperature. This is not always the case and less benign effects also occur.

Although in the troposphere both the air temperature and pressure for the most part fall steadily as one goes higher, anomalies do occur associated with weather. Sometimes, an inversion occurs when the normal decrease of atmospheric temperature or water vapour with height is reversed over a short vertical distance. If strong enough the inversion layer is capable of reflecting radio waves, particularly in the VHF and the lower UHF bands (but occasionally at the high end of the HF band) (Fig. 2.4). When this happens a **surface duct** is said to have formed. This results in anomalous long-range propagation by means of a sky wave. Almost all VHF/UHF terrestrial propagation over long distances is attributable to ducting.

As well as causing refraction, the presence of the atmosphere can also lead to absorption, although the effect is important only in the EHF band (>30 GHz) where it leads to so-called **absorption bands** and **windows** both of which have their uses. Absorption is due to

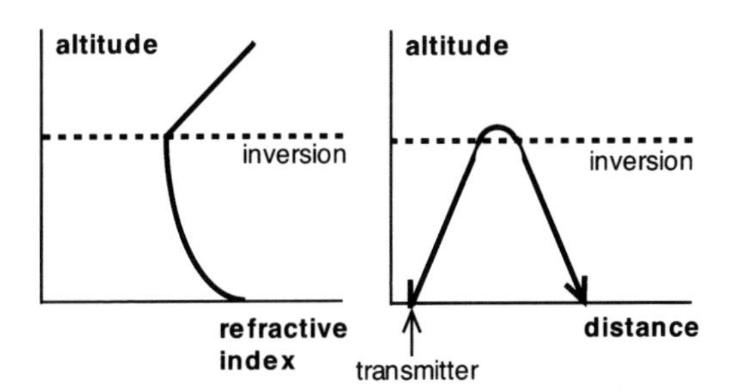

Fig. 2.4
The effect of temperature inversion on the refractive index of the atmosphere: a sky wave is returned to Earth by this surface duct.

collisions between radio quanta and gas molecules, in which the photons give up their energy. The quantum states of molecules are fixed by their internal structure, and are thus different for each different molecule. As a result each type of matter through which radio quanta pass, such as the gases of the atmosphere, has its own distinctive absorption spectrum, characterized by absorption peaks where the frequencies are just right to hit one of the maxima of collision cross-section.

A large absorption peak occurs at 60 GHz (see Fig. 1.3) and another at 120 GHz; these are due to collisions of radio photons with oxygen molecules. At sea level these peaks do not vary in size, and are independent of geographical location or weather. Peaks at a little over 20 GHz and just under 200 GHz are due to water vapour, as also is the rising 'floor' of the absorption curve, which in fact is the skirt of a very large peak at much higher frequency. These features are a consequence of the water content in the air, here assumed to be about the value for a dry day in Western Europe. However, the water content is heavily dependent on both the temperature and the weather conditions. It will be far more in a wet tropical environment and far less in an arid desert. In consequence both the 'floor' of the absorption curve and the water peaks will be very dependent on weather and location.

Use of the 60 GHz absorption band has been widely advocated as a way of improving the ratio of service to interference range for a radio transmitter. Because the signal falls at 14 dB per km in addition to the usual inverse square effect, it rapidly becomes negligible beyond the service range. The signal loss due to increasing range, expressed in decibels, can be written as the sum of two components:

$$L = L_{\text{spread}} + L_{\text{absoption}} \qquad (2.4)$$

The spreading loss term can be obtained from equation (2.3) above, hence we may write:

$$L = 10\log_{10}\left(\frac{A_{\text{r}}}{4\pi r^2}\right) + \beta \cdot r \qquad (2.5)$$

Here β is the absorption coefficient, and has the value 14 dB/km at the first oxygen absorption peak. Although high, this figure is invariant, so a fairly short link (say, less than 7 km) can be engineered to operate reliably. However, due to absorption alone the EIRP would have to be increased by a little over 14 dB (nearly thirty times) for every additional kilometre of range required, with a smaller increase also due to the usual spreading factor. In practice, therefore, the signal range is sharply limited. Probably the main commercial interest in the 60 GHz band is for short links where the attenuation can be tolerated and the negligible remote interference capability means very wide bands of frequencies can be assigned in particular locations.

Also of interest are the '**windows**', relatively low-absorption regions in the EHF band. However, such windows have a 'floor' attenuation determined by residual water vapour absorption, and therefore strongly dependent on weather conditions. The 29 GHz window extends from about 29 to 38 GHz and is used for both point–point communication and radar. As links are rarely more than 10 km long, the attenuation of 0.2 dB/km in the window will contribute only up to 2 dB of excess loss, which is acceptable, but this is only true in dry conditions, and things get measurably worse in rain of more than 0.5 mm/hour. In significant rainfall the attenuation increases sharply, roughly proportionally to the rate of precipitation. At 5 mm/hour (light rain) the excess attenuation will be 2 dB/km. Another significant window occurs at 95 GHz, although the attenuation here is four or five times larger (in decibels), still further reducing the range, which in this case is also very weather dependent.

However, note that the atmosphere rapidly thins with altitude (half is gone 5.6 km above the Earth's surface) so the atmospheric absorption of radio quanta also falls quite fast with height. The integrated atmospheric attenuation is small – even at its worst in the 60 GHz absorption band, a vertical transmission would suffer only 112 dB attenuation from the whole of the Earth's atmosphere, while at other frequencies the loss is very much less. For this reason the use of the EHF band is practicable for satellite communication, and is increasingly seen as attractive.

2.4 At ground level

It is terrestrial communication at or near ground level which is of the greatest commercial significance because human populations are concentrated there. The effects of the atmosphere on the passage of radio photons is significant in this case, but even more important is the influence of solid objects, the largest of which is the Earth itself. These lead to **reflection**, **shadowing** and **diffraction** effects which greatly complicate the prediction of received radio signals.

In fact the near-Earth terrestrial environment is usually full of reflectors. From VHF to SHF most building materials are partial reflectors of radio quanta, while metals are almost wholly so. Materials which reflect poorly, such as glass, are often backed up (in glass clad buildings) with other materials (metal, building blocks) which reflect better. Thus virtually all normal buildings and almost all road vehicles can be treated as good reflectors of radio energy in these bands. So also can hills, mountains and other terrain features. Above all, the surface of the Earth itself reflects at all frequencies down to the lowest, due to its large size.

So far as radio propagation is concerned, the most dramatic effects are caused by specular reflection, which is very common. For photons to be reflected the objects acting as 'mirrors' will generally have dimensions of the order of at least a wavelength and usually much larger, although strong reflections can occur from smaller objects which happen to be near their resonant frequency. However, resonant reflectors are rare at frequencies in the MF and lower, due to the scarcity of large-enough conducting objects.

The commonest non-resonant reflectors in the radio bands can be summarized as:

- ELF, VLF: Earth's surface (but deep penetration), ionosphere
- LF, MF, HF: Earth's surface, ionosphere
- VHF, UHF: Earth's surface, mountains, buildings, vehicles, ducting and ionospheric effects, meteor trails
- SHF: Earth's surface, mountains, buildings, vehicles, trees, people, street furniture
- EHF: just about everything, excluding only upper atmosphere effects

Changes from band to band are not as sharp as this list suggests, instead there is considerable overlap at transitions.

As an example of the effects of radio reflection by the environment, consider the case of two hill-top radio stations that communicate over a flat plain. There are two paths taken by the radio quanta travelling from the transmitter to the receiver: the direct path, and a ground-reflected route. This is the simplest example of **multipath propagation**, one of the most important phenomena of radio systems, and likely to occur in all radio systems operating other than in free space. Multipath phenomena have two marked effects:

- Variation in the level of total received RF power, due to the phase relationships between the received signals coming by alternative paths. This is sometimes called fading, and is particularly noticeable in mobile systems where the receiving antenna may pass through maxima and minima as it moves. Because the path length difference is dependent on the position of the receiving antenna, total received signal becomes very sensitive to the spatial location of the receiving antenna, which is in effect moving through a pattern of

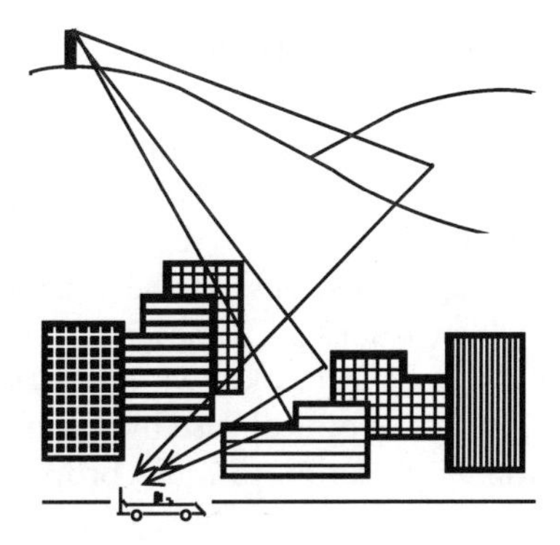

Fig. 2.5
The reflection from the hill would be strongest and arrive last.

interference fringes. Reflecting surfaces may be horizontal or vertical, or any other angle. The worst case is that of a reflecting surface on the line from transmitter to receiver and further from it. In this situation if the receiving antenna moves along this line, say in the direction of the reflecting surface, the direct and reflected paths will change in equal amounts but opposite sign. As the receiving antenna approaches the antenna the direct path length from the transmitter will increase by exactly the amount that the reflected path length decreases. Thus for a movement of $\lambda/4$ the two received signals will change from being (for example) in phase to being in antiphase, and thus the resultant will change from a maximum to a minimum (zero if the two component signals are of equal strength). To put it another way, minima will be spaced $\lambda/2$ apart.

- Confusion between modulation carried by the signals arriving over the shortest paths and those which are more delayed (Fig. 2.5). This could produce slight echo effects on analogue voice transmissions, but much more importantly can also give rise to confusion between successive digits in a digital transmission, particularly at high bit rates. If the bit rate is B b/s, evidently errors are probable for all path length differences $\Delta \sim c/B$ or greater. Thus if the bit rate B is n Mb/s the critical path difference is $300/n$ metres, which can easily arise for $n > 1$.

2.5 Shadowing and diffraction

Solid objects in the radio propagation environment not only reflect radio quanta but also obstruct and absorb them, giving rise to radio shadows (Fig. 2.6). However, the region of the geometrical shadow is not entirely free of photons; at the edge of the obstruction an effect occurs which results in some energy being thrown into the shadow region. This effect is known as **diffraction**. Its theory was first worked out for the optical case by Augustin Fresnel. His ideas apply with little modification to radio propagation.

For historical reasons diffraction phenomena are classified into two types: Fraunhofer and Fresnel diffraction. Fraunhofer diffraction treats cases where the source of light and the screen on which the pattern is observed are effectively at infinite distances from the intervening aperture. Thus, beams of light are parallel, so the wavefront is plane, and the mathematical treatment of this type of diffraction is reasonably simple. Fresnel diffraction, rarely encountered in radio propagation, treats cases in which the source or the screen are at finite distances and therefore the light is divergent.

A simple mathematical treatment of Fraunhofer diffraction will demonstrate the characteristics of radio diffraction at a knife edge, which is of most practical significance. The edge of any

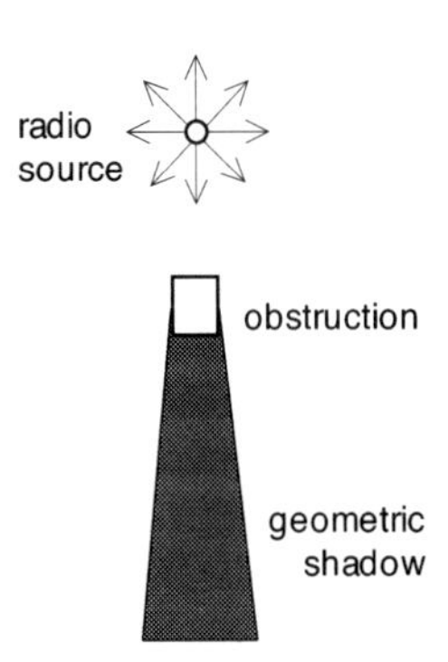

Fig. 2.6
Solid objects can absorb or reflect quanta to create radio 'shadows'.

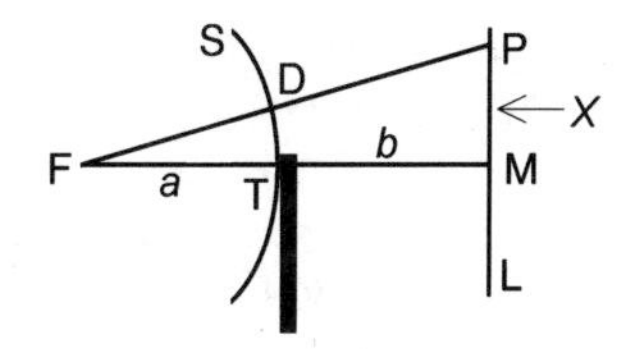

Fig. 2.7
Diffraction at a knife edge.

solid object approximates to it, and it is even quite a good model for the case of a radio transmission passing over a ridge in the terrain.

Consider a flow of radio quanta towards the knife edge; for simplicity of analysis we assume a cylindrical front S (Fig. 2.7). How does the density of photons vary in the plane PML? The point P is x above M and we calculate the flow of quanta to this point.

The front can be considered as a series of horizontal strips, known as Fresnel zones, which are parallel to the knife edge, each strip having a lower boundary line half a wavelength further from P (Fig. 2.8). If the sth zone is distant r_s from P at its lower edge, the $(s+1)$th is distant $r_{(s+1)}$ and so on, then:

$$\left[r_{s+1} = r_s + \frac{\lambda}{2}\right]_{\text{all } s} \tag{2.6}$$

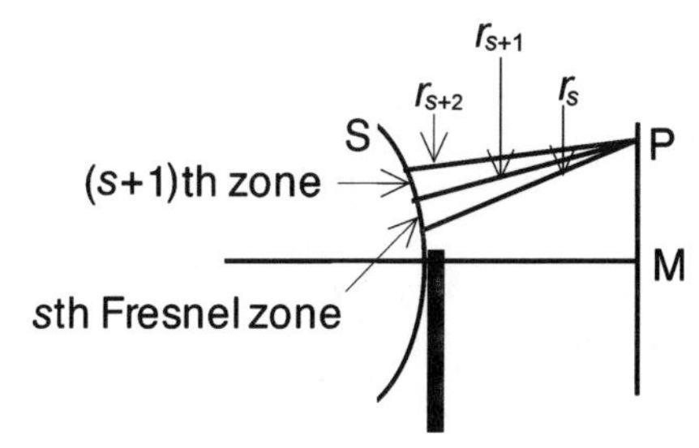

Fig. 2.8
The Fresnel zones.

Thus every Fresnel zone is of approximately the same area, so that the same number of photons is passing through it, and each is half a wavelength further than its neighbour from P. If we consider the wave functions of the individual quanta in each Fresnel zone, clearly each individual one will have a neighbour in the next zone which is further from P by a half wavelength, and thus the wave functions of each such pair will arrive at P in antiphase, and so will cancel. (The difference in distance from P between adjacent pairs of zones is very small, therefore any difference of photon density due to different spreading may be neglected.) It follows that if the number of Fresnel zones in the wave front is even the wave intensity at P will be zero, but if odd it will be a maximum.

The series of zones on the side DS is unobstructed, so it will give half the amplitude of the whole wavefront. On the side DT the wave is incomplete, being partly blocked off by the knife edge. As above, if there is an odd number of zones the result at P is a maximum, and if even then it is a minimum.

Assuming x small compared with b:

$$\mathrm{PT} = \sqrt{b^2 + x^2} \approx b + \frac{x^2}{2b} \tag{2.7}$$

similarly:

$$\mathrm{PF} \approx a + b + \frac{x^2}{2(a+b)} \tag{2.8}$$

so:

$$\mathrm{PD} = \mathrm{PF} - a \approx b + \frac{x^2}{2(a+b)} \tag{2.9}$$

For a maximum at P, PT − PD $= n\lambda$ (resulting in an odd number of zones) which gives:

$$\frac{x^2}{2b} - \frac{x^2}{2(a+b)} = n\lambda$$

so for a maximum:

$$x_{\max} = \sqrt{\frac{2n\lambda b}{a}(a+b)} \tag{2.10}$$

Similarly for a minimum:

$$x_{\min} = \sqrt{\frac{(2n+1)\lambda b}{a}(a+b)} \tag{2.11}$$

Thus if, for example, $b = 10\,\text{m}$, $a = 10\,\text{km}$ and $\lambda = 1\,\text{m}$, maxima occur at $x = 4.47\sqrt{n}$ metres, that is 4.47, 6.32, 7.5 metres and so on, with minima between them.

If P is below M all the zones are obscured on one side of D, and the first on the other side starts some distance from D. Thus the intensity diminishes steadily from M to zero over a little distance.

Note that the shorter the wavelength the smaller the dimensions of the diffraction patterns. Thus the wavelength is a scaling term throughout; if the wavelength is reduced by a certain factor all the diffraction effects will physically scale down in size by the same factor. Writing all the dimensions as numbers of wavelengths we have

$$x'_{\max} = \sqrt{\cdot \frac{2nb'}{a'}(a'+b')} \tag{2.12}$$

where $x'_{\max} = x_{\max}/\lambda$, $a' = a/\lambda$ and $b' = b/\lambda$.

It is the position relative to the edge of objects *measured in wavelengths* which determines the diffraction effects, so the size of diffraction 'fringes' around objects depend on the wavelength.

If a radio link is obstructed by a terrain feature (such as a ridge, treated as a 'knife edge') the signal does not fall at once to zero in the geometric shadow, nor does it immediately reach a uniform maximum value out of the shadow, as is evident from the preceding analysis. It is clear that if the wave path is above the knife edge there will still be possible signal variation due to diffraction, although this effect will diminish rapidly the higher the path. If the nominal signal path is below the knife edge the signal will be heavily attenuated but not zero, and the reduced number of quanta that do get through may still be used for communication.

The practical consequences of diffraction for radio propagation depend on the size of the obstacles relative to the wavelength of the quanta concerned. There is diffraction even around major terrain features such as mountains by ELF, VLF and LF waves, with wavelengths measured in many kilometres, but shadowing by terrain features is increasingly severe for radio transmissions with shorter waves than a few hundred metres length, and hence in the MF, HF and higher frequency bands.

But what of manufactured objects? The characteristic dimension of human beings is 1.5 to 2 metres and very few of their open-air artefacts are more than an order of magnitude different from this, that is most are less than 20 metres in dimension but more than 0.15 metres. In a flat-terrain built environment (suburbs, city) there will consequently be little shadowing of waves longer than about 10 metres ($f < 30$ MHz), since diffraction around all objects will largely fill in shadows. Thus in the HF bands and below (in frequency) shadowing by built structures is unlikely. At VHF there will be serious shadowing only by substantial buildings and the largest vehicles, but still with significant signal received some metres within the geometrical shadow area. At UHF smaller objects, ordinary cars, road signs and other street furniture, will create shadows and it will be necessary to approach within the order of a metre of the geometrical shadow edge to receive signals. The effect grows even more marked as the wavelength gets shorter, until with millimetre waves (EHF) virtually all radio-opaque objects in the human size range throw deep and sharp-edged radio shadows.

2.6 Radio propagation in a complex built environment

As already indicated, in all bands up to HF there is good penetration of radio waves into the built environment because the waves are large enough to diffract round most of the objects encountered. At VHF and above, although diffraction plays a significant role in radio propagation in the typically cluttered urban or suburban radio environment, reducing the obscuration by objects not too large compared with the wavelength, an even more important part is played by reflection. When there are many reflecting bodies in close proximity to the receiving antenna and no line of sight (or near line of sight) path exists, radio quanta are propagated entirely by reflection from buildings, and to a lesser degree vehicles, assisted by diffraction round them. This **scattering propagation** scenario is of great importance. In the near-to-ground environment on Earth it is the dominant mode of propagation at VHF, UHF and above.

A full analysis of scattering propagation is inevitably complex. The pioneer contribution to the subject was by J. J. Egli as early as 1945, and the paper by Allsebrook and Parsons (1977) is deservedly regarded as a classic. The theory is extensive, and only the principal results will be summarized here. First we consider the case of transmissions in the form of unmodulated carriers, so that only amplitude and phase variations are significant, and subsequently the case of pulse modulation will be reviewed, to bring out the time-domain significance of scattering propagation.

As might be expected, this type of radio propagation, which relies on the chance that photons will be reflected by objects which happen to be in the right location to bounce them to the receiving antenna, cannot easily be described by solutions of Maxwell's equations. Although calculations can be performed for any particular configuration of transmitter, scattering sites and receiver, each is different from all others. However, when there is (a) no line of sight between the transmitter and receiver and (b) many significant scatterers (at least five or six within one hundred wavelengths) the propagation converges to a stable statistical model, sometimes referred to as **Rayleigh fading**, with characteristic both local and longer range signal variation.

That there should be no line of sight path for quanta is highly likely in most built-up environments, at least for receiving antennas at street level or not much above. Experimental studies in cities show

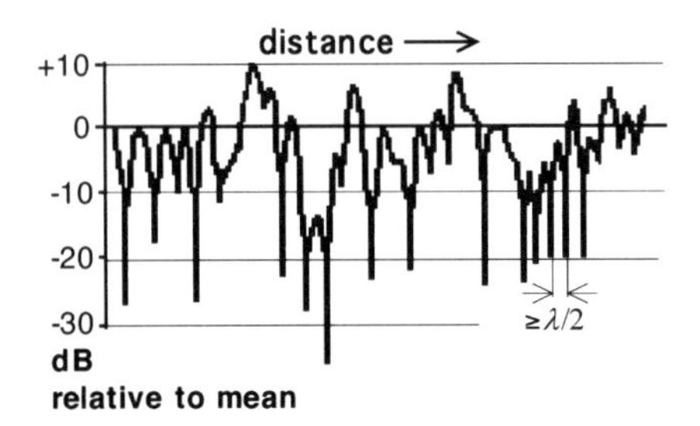

Fig. 2.9
How a Rayleigh distributed signal varies in amplitude with position of the receiving antenna.

that the number of scatterers is Gaussianly distributed with a mean number of 32 and a standard deviation of 12. Evidently the conditions necessary for Rayleigh signal statistics will be met very commonly.

What does the scattered radio signal look like? Over variations in position of up to a hundred wavelengths the signal is characterized by a Rayleigh distribution of amplitude and random phase (Fig. 2.9). Longer range signal variation may be characterized as a slow variation of the mean of the Rayleigh statistical distribution which represents the short range signal variation. This is the key to the satisfactory description of scattered radio transmissions.

First the Rayleigh distribution, for local variations. Where all propagation is by scattering, the probability density function $p(e)$ of amplitude e of received signal is:

$$p(e) = \frac{e}{\sigma^2} \exp\left(\frac{-e^2}{2\sigma^2}\right) \tag{2.13}$$

where σ is the modal value of e.

This is a very highly variable distribution (Fig. 2.10). For comparison, after being filtered into a narrow bandwidth, Gaussian noise also has a Rayleigh amplitude distribution. Thus it is not unfair to say that the amplitude distribution of the received scattered signal is about as random as it could possibly be.

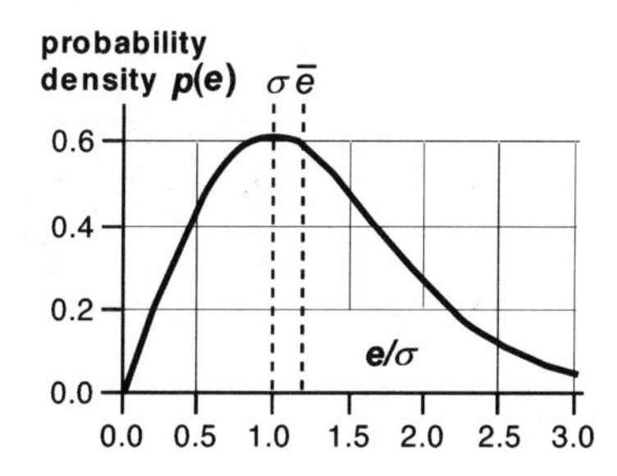

Fig. 2.10
The Rayleigh distribution.

The probability that the received signal e is greater than some value E_{min} is (by integration over the limits 0 to E_{min}):

$$p(e \geq E_{min}) = \exp\left(\frac{-E_{min}^2}{2\sigma^2}\right) \tag{2.14}$$

Putting $p = 0.5$ in the above expression, it is easy to obtain the mean value of e as 1.18σ (+1.5 dB relative to s, bearing in mind that this is a voltage ratio). Also this cumulative distribution permits the calculation of the probability that the signal will exceed some minimum value (E_{min}) needed to give the required received **signal-to-noise** (S/N) ratio. For 99% probability that the signal exceeds a value E_{min} the value of σ is 17 dB greater than E_{min}. This is the reason why designers often set a mean signal 20 ($\approx 17 + 1.5$) dB above the minimum acceptable level as a 'safety margin'; it means that the signal will fall below the desired minimum in less than 1% of locations.

More generally, the probability that the signal will be of inadequate magnitude to give satisfactory communication is just one minus the probability calculated from equation (2.14). In Fig. 2.11 this failure probability is plotted against the number of decibels by which the mean signal exceeds the threshold of the receiver. A mobile system user will experience irregular breaks in reception due to the signal falling below the receiver threshold in the troughs of Rayleigh fades. The graph gives the probability that this will happen, and therefore the proportion of the time that there will be no reception

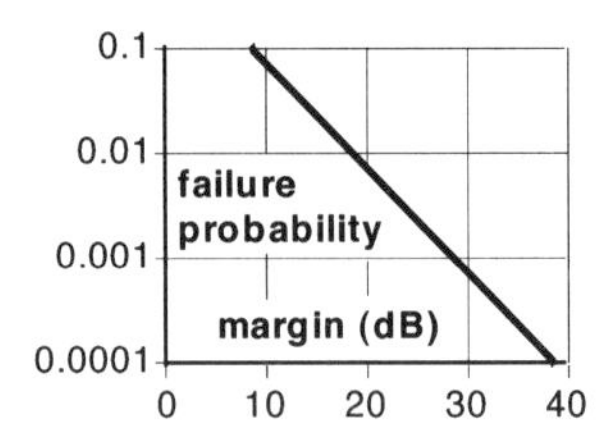

Fig. 2.11
Probability of failure to communicate as a function of 'safety margin'.

for a user who continues in motion. In a digital system it represents the minimum 'floor' below which the raw bit-error rate cannot fall. For a stationary user the same figure is simply the probability that it will prove impossible to establish a radio connection from the particular location being occupied. So, for example, if a mobile phone user attempts to make a call, Fig. 2.11 gives the probability of being unable to do so. If the probability is 0.01, at a margin of 18.5 dB, then one call in every hundred attempted from randomly chosen locations will be frustrated. The probability gets an order of magnitude less for every 10 dB increase in mean signal.

As to the phase of the received signal, in the Rayleigh distribution all phases are equiprobable, so if ϕ is the phase its probability density function is:

$$p(\phi) = \frac{1}{2\pi} \tag{2.15}$$

where ϕ is in radians.

Sharp phase changes occur in fading troughs, resulting in spurious outputs in phase or frequency modulated systems.

2.7 Longer range signal variation in the near-to-ground environment

Within a few tens of wavelengths of a fixed point the mean value of the received signal is approximately constant, so the Rayleigh distribution is a good approximation to the distribution of signal amplitude. Over longer ranges the variation of signal may be represented as a variation of mean value and hence σ. Obviously the mean signal strength will tend to fall as the distance from the transmitter increases. As we have seen, in free space propagation an inverse square law applies but under conditions of scattering propagation the mean signal falls more rapidly with range, as a consequence of multipath effects. Indeed, this is the case under multipath conditions less extreme than a true Rayleigh fading environment.

Further analysis leads to the conclusion that it is the inverse fourth power law which best relates both the Rayleigh distribution mean and modal values to range from the transmitter. This is in sharp contrast to the inverse square law which applies in space, and demonstrates that the range of a transmitter which is obliged to depend on scattering propagation will be much reduced.

How well is it borne out in practice? If it were exact the mean signal should fall by 40 dB for a 10 : 1 increase in range. This gradient has been measured, both for European and US cities, and results are mostly in the range 40 ± 3 dB, generally coming out on the high side in city centres and lower in suburban areas. However, there are some exceptions. At sidewalk level in downtown Manhattan, surrounded by closely spaced and exceptionally high buildings, a figure of 48 dB has been recorded. By contrast central Tokyo shows the unusually low figure of 31 dB, as a result of the low-rise buildings and the topography of the city, surrounded by hills on which the radio transmitters are sited, so that line-of-sight transmissions are not infrequent. Even so, substantial deviation from the 40 dB figure is exceptional, and the inverse fourth power law remains a useful indicator of likely system performance in the absence of measured field statistics.

By using the inverse fourth power to model the variation in the value of σ in a Rayleigh distribution it is a straightforward matter to predict the performance of terrestrial radio systems, particularly area coverage and co-channel interference (Gosling 1978).

It is also important to consider the effects of scattering propagation in the time domain. This has been studied experimentally, particularly in the VHF and UHF bands, by using short pulse transmissions (or their equivalent) and investigating the relative amplitudes and time of arrival of pulses at the receiver. As would be expected, the amplitude depends primarily on the reflecting area, while time of arrival is proportional to total path length, 3.3 μs for each kilometre. The first pulse to arrive may therefore not be the largest, since it may come from a relatively small close-in reflector.

Many studies have been reported of the spread of arrival amplitudes and times, which are obviously critically dependent on the density of scatterers in the physical environment of the receiving antenna. In the UHF band, for a flat rural area delay spreads of 0.1 to 0.2 μs have been reported, in suburban areas the likely spread is 0.5 to 1 μs, whilst in urban areas delay spreads of around 3 μs are commonplace, but values as high as 10 to 100 μs have been observed, though rarely. The figures quoted are not very frequency dependent until the millimetre wave bands are approached, when they tend to fall (nearer scatterers become more significant). Similarly larger values (more distant reflections) are recorded primarily at VHF.

As indicated in Section 2.4, delay spread becomes significant if the radio signals have digital modulation. If the delay spread is of the order of the inter-symbol interval severe increases in raw error rates occur (i.e. at about 300 kilobaud when the delay is 3 μs) because '0's and '1's received by different paths overlap. This effect may be usefully offset by the use of equalizers.

2.8 The effects of buildings

So much for radio services to the out-of-doors user. However, to give true universality of communications it is now increasingly required to be able to use portable radio equipment inside buildings. This involves two possible modes of operation: the reception of transmissions originating within the building and communication with stations outside, which involves penetration by the radio transmissions through the walls.

Building penetration by radio quanta has been most thoroughly studied at 800–900 MHz for cellular radio telephone services, with less intensive investigations at other frequencies. Needless to say, the means of building construction is critically important. An all-metal building with no windows hardly admits any radio energy at all, but fortunately except for military installations and semiconductor foundries very few buildings are constructed in this way.

Even so the techniques of building are dominant. For example, at 800 MHz the path loss difference inside and outside a building at first floor level in Tokyo has been measured as 26 dB, but similar studies in Chicago yield 15 dB. This is because typically much more metal (such as mesh) is used for building in earthquake-prone Japan than in Chicago, where earthquake risks are slight. In Los Angeles, where the earthquake risk is intermediate, the measured loss is 20 dB. Results of measurements are only meaningful, therefore, when related to building techniques.

Experimental studies in Europe show a penetration loss in the range from a few decibels up to 20 dB on the ground floor, tending to decrease with increasing frequency (presumably due to better penetration through windows and other apertures), and with loss decreasing on higher floors, typically by 2 dB per floor. Signals are Rayleigh distributed within the building. These conclusions have been confirmed for European building styles by several studies, and may be taken as valid for the VHF and UHF bands. In the SHF, particularly above 10 GHz, the transparency of buildings to radio signals decreases, and for millimetre waves (EHF) there is virtually no penetration.

Because of the interest in, for example, radio-based cordless PABXs (internal telephone systems), telephones and other cordless systems, the propagation of radio waves within buildings has also been widely investigated in recent years. Studies have been mostly conducted at UHF and above, because of the need for small antennas inside buildings, better penetration of apertures within buildings at shorter wavelengths and less restrictive frequency assignment availability from the radio regulatory authorities.

As might be expected, room-to-room propagation is by scattering, with the geometry such that in general all signals will have undergone several reflections before being received. This makes for a rapid rate of signal attenuation with distance, along with Rayleigh statistics of signal variation. Wavelengths should ideally be small compared with door and corridor dimensions, so frequencies below 300 MHz result in less uniform coverage. With low powers there is little energy radiated outside the building from transmitters within, which is good for spectrum conservation and yields a degree of privacy.

An indoor channel is found to have the advantage of relatively small delay spread (because of the close proximity of reflecting surfaces), limiting the undesirable effects of multipath transmission. Thus acceptable performance (raw BER $< 10^{-4}$) can usually be obtained without the need for channel equalization at bit rates up to 3 Mb/s, which allows the design of low cost user terminals for many applications.

These conclusions remain valid up to about 20 GHz. Above this frequency internal partitions become opaque and radio shadows are not significantly filled in by diffraction. Use of millimetre waves within buildings concentrates on covering each room by its own transmitter, located on the ceiling and transmitting down on to the user population, to put radio shadows in unimportant locations. If room-to-room communication is required, cable-linked active repeaters are used.

Questions

1. A transmitter radiating 1.0 W at 150 MHz has a main-lobe gain of 20 dB relative to isotropic and is oriented vertically toward a space probe at a range of 1000 km, which has a short dipole antenna optimally aligned to receive the signal. What is the received power? (−86 dBm)

2. A radio point–point link operates at 30 GHz over a distance of 2.86 km, and is both transmitted and received by a paraboloid antenna of 20 cm diameter. The operating frequency is changed to 60 GHz. With the same RF power to the transmitter antenna, what must be the diameter of the two paraboloids to give the same received signal power as at the lower frequency? (1 m)

3. A VHF community broadcasting station operates from a low antenna in a flat urban area. Its licence is modified to permit an increase in transmitter power from 50 to 150 W. By how much would you expect its service range to improve? (32%)

4. An urban mobile telephone system uses a low transmitting antenna. At 3 km from the transmitting site about 1% of randomly chosen locations fail to give an acceptable signal level. If the system specification allows failure at 10% of locations, what is the likely maximum range? (5.3 km)

5. What are the problems in using millimetre wavelength transmissions for urban radiotelephone systems?

CHAPTER 3

THE LONG HAUL

Although the most commercially important applications of radio technology at present are for short- and medium-range applications, from cellular phones to terrestrial broadcasting, long-haul communication systems remain important. It is true that much intercontinental traffic is handled by optical fibre, but there are many situations in which radio remains the carrier of choice. Mobile communications requirements must be met, for at least part of the transmission path, by using radio. But in addition there can be compelling reasons for using radio even where the stations are at a fixed location. Obviously this is true in more remote places where there is no access to the optical fibre network, also where temporary communications must be set up and broken down quickly as on civil engineering projects and not least where the physical security of the link must be high enough to survive even disaster situations. These considerations dictate radio solutions in many civil situations, and often dominate emergency service and military thinking.

3.1 Surface effects

For radio links up to 100 kilometres or so the technology already described is adequate, but proximity to the Earth's surface does lead to increasingly severe losses over longer distances, which is a problem for the long-haul communicator. On any large conducting surface, such as the Earth, one possibility is to propagate radio quanta characterized by surface waves. The fields associated with the passing quanta induce currents in the conducting plane (the Earth's surface) which play a critical part in the propagation mechanism. For VLF to MF frequencies this is the only practicable transmission mode, since to be even approximately clear of Earth effects it would be necessary to elevate antennas by several wavelengths, which is not practicable.

Surface waves are launched by vertically polarized antennas over the Earth's surface. The Earth is spherical, so propagation over this curved surface depends upon diffraction. This mode is characterized by excellent phase and amplitude stability at the receiver, due to the simple propagation mechanism which is independent of almost all variable physical factors. The mathematics of this mode of propagation is complicated, and leads to a model, with a perfectly conducting earth, in which received power falls inversely with the square of distance. In practice the earth is not perfectly conducting at these frequencies and the induced currents dissipate energy, so the rate of fall of signal is much greater, particularly at MF where it approaches an inverse fourth power.

VLF is limited to slow data rate transmissions because of the limited bandwidth, but they can be received worldwide. Using very high power for hand-speed Morse code transmissions, from 1910 to 1927 or so this was the only intercontinental radio communications technology, but with the introduction of HF radio in the mid-1920s a much cheaper alternative became available, allowing telephony as well as telegraphy. Surface wave VLF continues in use today where its unique characteristics of extreme stability and reliability are valued. It has military interest for communicating with submerged submarines.

The surface propagation mode is also the basis of MF and LF broadcasting (up to at most 300 km range LF, 100 km MF). Because of the higher centre frequencies the bandwidths of equipment are proportionately higher than at VLF, which permits their use for sound broadcasting. The surface wave propagation mode is used to a very limited degree at HF for short-range (a few kilometres) military and rural radiotelephone services, using vertical whip antennas. At VHF and above the attainable range is too small to be of use.

3.2 Avoiding surface losses

Losses over the Earth's surface are increasingly severe as frequency is increased. The only way to escape from this limitation is to direct the radio energy away from the Earth into space, where losses are much lower provided that we stay below frequencies where atmospheric absorption is significant. This is **sky-wave** communication; its essence consists of directing the main lobe of the transmitting antenna well above the horizon, so that the quanta are launched on a path away from the Earth's surface, and then using some means to return the radio energy back to the ground and close to the intended receiving station. The sky wave has the usual inverse square spreading loss with increasing range, and there may in some cases be an additional loss when the wave is redirected back to Earth, though this is not always so.

What of the atmosphere itself? Can collisions between photons and air molecules deflect some of the radio energy back to the ground? Although the troposphere is normally regarded as transparent to electromagnetic radiation up to around 20 GHz, at frequencies between 1 and 5 GHz in the troposphere there is sufficient Thomson scattering of radio quanta to make propagation by this means possible. This **tropospheric scatter** consequently provides a means of long distance (100–1000 km) communication, but with considerable path loss (>140 dB). The technology depends on transmitting very high powers from high gain antennas, launching transmissions at low angle ($<5^{\circ}$) to equally high gain receiving antennas. Although 'troposcatter' can give reliable links, it has now been eclipsed by other communication technologies.

We have already seen that ducts, caused by inversion layers, are naturally occurring atmospheric phenomena which will redirect sky waves back to Earth for transmissions in the VHF and lower UHF bands. However, these are chance meteorological phenomena which do not give reliable means of communication (other than in a very few exceptional locations). Much more certain methods of securing the return of the sky-wave radio transmissions to Earth exist and form the basis of several radio communications technologies currently in widespread use.

Current sky-wave communication systems use one of three techniques for returning the signals to earth: scatter of radio quanta by **meteor trails**, propagation by reflection of quanta by the **ionosphere**, and artificial Earth **satellites** carrying radio receivers and transmitters, and capable of giving a large power gain between the received and retransmitted signals.

3.3 Meteor scatter

A meteor, often called a shooting or falling star, is seen as the streak of light across the night sky produced by the vaporization of interplanetary particles as they enter the atmosphere. They are the debris from comets, which leave a trail of matter behind them as they orbit the Sun. As the Earth passes through this cloud of debris, many particles, on average about the size of a grain of sand, enter the atmosphere at speeds up to 100 km/s. Air friction causes them to vaporize, creating a large elongated trail of very hot gas, the visible shooting star, in which the temperature is high enough for the outer electrons to be stripped off atoms in the atmosphere, leaving clouds of positive ions and free electrons. The process of heating begins some 50 km out and the burning up is almost always complete by the tropopause, at 20 km.

The very large cloud of ionized gas produced is able to refract and reflect radio quanta, mostly because the free electrons have a high cross-section for collision with radio quanta in the VHF band (Fig. 3.1). However, the positive ions and free electrons readily recombine, so the phenomenon does not last; as soon as it is created the ionization begins to die away. This recombination happens faster the higher the atmospheric pressure, because high pressure brings ions and electrons closer together on average. So trails formed at higher altitudes will last longer than lower ones, although this is somewhat balanced by the trails at lower altitudes (where heating is fiercer) being more intense initially. Meteor-scatter radio

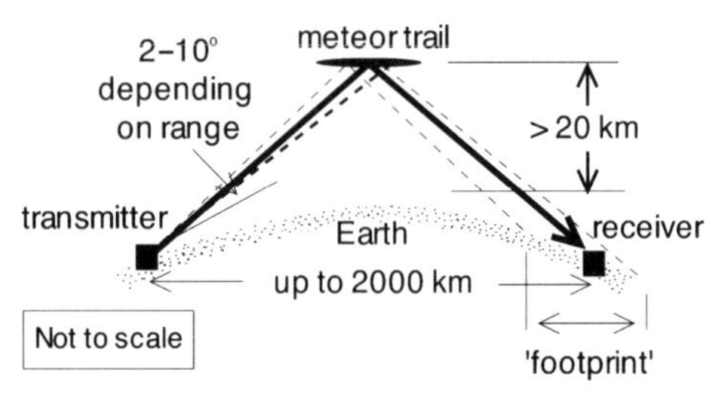

Fig. 3.1
Meteor scatter propagation.

systems work by directing a radio sky wave at a meteor trail, which reflects it back to Earth.

Because they are launched towards the reflector over a range of angles, the returning radio quanta are spread out over a '**footprint**' on the Earth's surface around the designated target point. Due to the geometry, at ground level there is nothing to be intercepted in the **skip area** between the transmitter and the 'footprint' where the signal returns to Earth. In consequence meteor scatter has good privacy (attractive to the military) and gives rise to minimal spectrum pollution because it cannot cause interference outside the 'footprint'. Systems operate over paths of up to 2000 km length, beyond which the geometry of the system dictates so low an angle of elevation for the transmission that losses due to the surface and terrain features become excessive, although hill-top sites for transmitter and receiver can help. If the link is to be used exclusively for long range the antenna main lobe width can be very much smaller, with consequent improvement in gain and usable trail duration. Systems designed to cover many differing transmission paths must have a wider lobe.

The principal difficulty of meteor-scatter communication is that the transmission path, although utterly reliable in the long run, is only present intermittently. Over a given path the probability density function for a wait t before a path is open is $p(t)$, where approximately:

$$p(t) = \frac{1}{\tau} \exp\left(\frac{-t}{\tau}\right) \tag{3.1}$$

The probability of a channel appearing in a time T is thus:

$$P(T) = \int_0^T p(t) \cdot \mathrm{d}t = \left[-\exp\left(\frac{-t}{\tau}\right) \right]_0^T = 1 - \exp\left(\frac{-T}{\tau}\right) \quad (3.2)$$

In middle latitudes the value of the average wait is typically 2.5 minutes. Once a path is open there is a similar exponential distribution of useful channel 'life', that is the time until the S/N ratio has fallen too low to sustain the required data transmission rate. Obviously this is longer the higher the transmitter power and the lower the minimum acceptable signal level. However, because the ionization decays exponentially raising transmitter power substantially only gives a modest extension of the 'window' duration, and the typical transmitter power (100 to 500 watts) is mostly determined by what is available at moderate cost. Higher altitude trails recombine more slowly, and so last longer, but they are less frequent. With all the variable factors, a typical mean for the usable channel duration is only 0.5 second.

Even so, whilst a channel is active data can be passed at a high rate (up to at least 100 kb/s, depending on hardware specifications) but the short duration of the functioning life of each meteor trail means that the average message capacity over a 24-hour period is commonly no more than 200 b/s in each direction. However, even this amounts to 1.6 Mb per day in each direction, which is enough for many e-mail, data monitoring and short messaging requirements.

Meteor-scatter systems are an example of burst transmission 'store and forward' operation, in which digital traffic is stored until the channel opens and is then rapidly forwarded to its destination. Meteor scatter only achieves a modest average transmission rate, but has a number of advantages which result in its use in niche applications to which these particularly apply. Its most important virtue is reliability: the channel will always be there, sooner or later, because new meteorites are always entering the atmosphere, many thousands every day. This is not affected by ionospheric conditions, weather and so on, and there are only moderate changes from time to time, as the Earth passes through meteor showers. It is also cheap. Used remote from telecommunication services, particularly in underdeveloped areas, meteor scatter is valued wherever a moderate data transmission rate with delays of up to a few minutes is acceptable, for example in meteorological, water resource and environmental management systems.

3.4 Ionospheric propagation

Although its existence had long been suspected, notably by Nikola Tesla (1856–1943) (who unsuccessfully built a tower to try to make contact with it), Edward Appleton first obtained observational evidence for the ionosphere in 1925. The ionosphere is a consequence of the ionization of the upper atmosphere by energetic electromagnetic quanta emitted from the Sun, mostly ultraviolet but also soft X-rays. These quanta collide with air molecules or atoms and strip off their outer electrons, leaving them positively ionized in a sea of free electrons. However, at a rate depending on local atmospheric pressure, the process of recombination is going on all the time, so that the degree of ionization is a consequence of a dynamic equilibrium between continuing ionization and recombination. It is therefore greatest over a particular part of the Earth's surface in summer and daytime, least in winter and at night.

The ionosphere (Fig. 3.2) is subdivided into:

- the D layer between 60 and 85 km in altitude, which disappears at night due to the rapid recombination in the relatively high pressure
- the E layer (sometimes called the Heaviside layer) which is at an altitude between 85 and 140 km
- the F1 and F2 layers which are found above 140 km and merge at night (they were formerly jointly called the Appleton layer)

In the D and E layers it is air molecules which are ionized, but in the F layers pressure is so low that atmospheric gases exist primarily in atomic form, so that the positive ions are atoms. The ionized gases

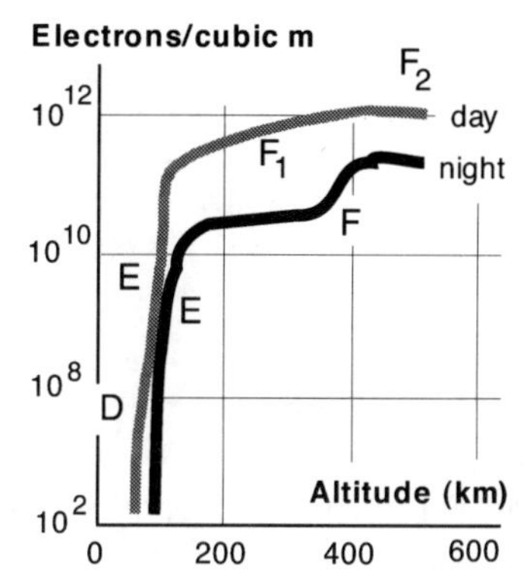

Fig. 3.2
Typical variation of free electron density with altitude by day and night.

in all these layers can refract and reflect radio quanta in just the same way as described above for meteor trails.

The ionosphere has two effects on radio quanta: absorption and refraction. By collisions with the radio quanta, free electrons receive energy which is subsequently dissipated in further collisions with electrons, positive ions or gas atoms. As a result energy is lost from a radio transmission passing through a region of high density of free electrons. The cross-section of an electron for collision with a photon, and hence the probability of such an event, depends on the energy of the photon. Alternatively the free electron, which has been raised to a higher energy state by the colliding photon, may relax back into its initial energy state, before it has had time to give up its energy by further collisions, releasing a new photon with energy identical to the original. The effect is exactly equivalent to a reflection of the photon. The path of the stream of radio quanta will consequently be redirected, and the result can be that the radio transmission is turned back towards the Earth, as a consequence of this collision process. Thus, in the ionosphere photons will sometimes escape, if their energy is high enough or the free electron density low enough, and otherwise they may be returned to Earth. What the outcome will be depends on the electron cross-section for collision, the length of the path in the ionized layer and the density

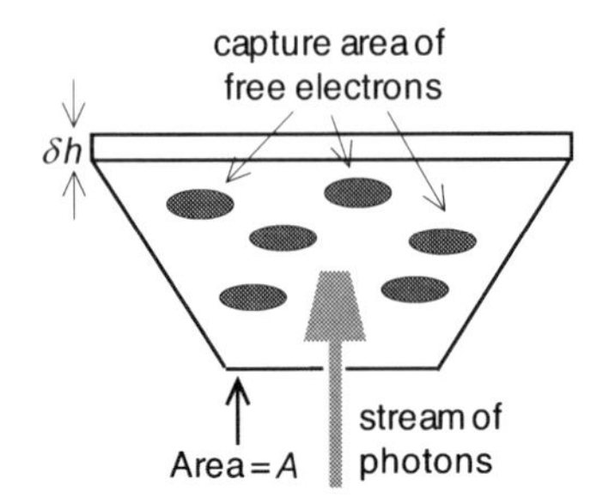

Fig. 3.3
A radio transmission comprising a stream of radio photons traverses a thin slice of an ionized layer.

of free electrons, factors which jointly determine how many collisions are likely to take place.

We begin the analysis with the case of a radio transmission directed vertically upward at an ionized layer (Fig. 3.3). A stream of radio quanta passes through the ionized region; consider a thin slice of area A. The probability that there will be a collision depends on the cross-section for collision of the free electrons. What is this cross-section for collision? Simply the area within which a passing radio quantum will be captured by ('collide with') a free electron. Just as in receiving antennas, which also capture quanta, what we are considering here is just the capture area (or aperture, in antenna language) of an individual electron. This area is therefore (as with an antenna) of the form $k\lambda^2$ where k is some constant. If the density of free electrons is ρ, the number in a horizontal slice of the ionosphere is $\rho \cdot A \cdot \delta h$ so the probability of a collision in the slice is:

$$p = \frac{k\lambda^2 \rho A \cdot \delta h}{A} = k\lambda^2 \rho \delta h \tag{3.3}$$

In the limiting case, the value of p must exceed a certain minimum in order that there shall be enough collisions to return significant numbers of quanta to Earth, so:

$$p \geq p_{\min} \tag{3.4}$$

Hence:

$$\lambda \geq \sqrt{\frac{p_{\min}}{k\rho \cdot \delta h}} \tag{3.5}$$

or, in frequency terms:

$$f \leq \sqrt{\frac{c^2 k \delta h}{p_{\min}} \cdot \rho} = f_c \tag{3.6}$$

Here f_c is the critical frequency, which is the frequency at which the return to Earth of the radio quanta just begins to fail. Clearly:

$$f_c \propto \sqrt{\rho} \tag{3.7}$$

The critical frequency for the various ionospheric layers increases steadily with the height of the layer, since the high layers have lower recombination, and hence higher electron densities. It shows marked diurnal variation, rising during the day, when electron density increases due to the Sun's action, and falling at night, when it falls due to recombination. Critical frequency is also dependent on the season, being highest in summer, and on the strength of the Sun's activity, which varies over an eleven year cycle.

When the signal is not launched vertically but at a lower angle, the quanta pass through a longer slant path in the ionized layer, increasing the chance of a collision within the path length. The escape frequency at this angle is consequently increased, leading to a **maximum usable frequency (muf)** for communication given by:

$$f_m = \frac{f_c}{\sin \theta} \tag{3.8}$$

But the launch angle required is determined by the communication path length (the lower, the further). If $2d$ is the great-circle distance between two sites (that is, the curved path measured over the Earth's surface) then if the radius of the Earth is a, the angle subtended at the centre of the Earth by the path is $2d/a$. Thus the midpoint apparent rise in the Earth's surface is:

$$s = a\left(1 - \cos\frac{d}{a}\right) \tag{3.9}$$

For the triangle formed by the propagation path to a reflecting layer at height h and the straight line between the sites, the base angles are:

$$\alpha = \tan^{-1}\left[\frac{h + a\left(1 - \cos\frac{d}{a}\right)}{a\sin\left(\frac{d}{a}\right)}\right] \tag{3.10}$$

and (neglecting any effect of lower atmosphere refraction) the launching elevation of the radio signal is (Fig. 3.4):

$$\theta = \alpha - \frac{d}{a} \tag{3.11}$$

In practice h can only be approximated, since it varies with time of day and the seasons, and anyway the lobes of HF antennas used for ionospheric communication are wide enough to accommodate a few

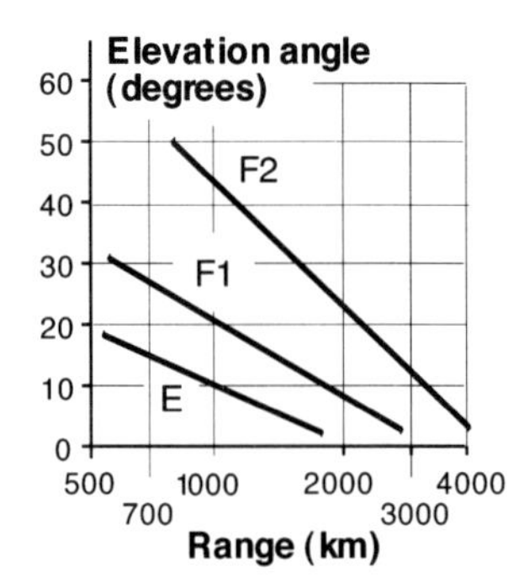

Fig. 3.4
Approximate antenna elevation for one-hop ionospheric propagation plotted against range.

degrees of error in elevation angle, so it is acceptable to use the simplified form:

$$\theta = \tan^{-1}\left(\frac{h}{d}\right) - \frac{d}{2a} \qquad (3.12)$$

However, there are limits on the usable angle of elevation from the ground. Less than 5° leads to large losses (as with surface waves) and the highest angle practicable with F layer reflection is about 74°.

This analysis assumes that the radio energy is projected into space by the transmitting antenna, reflected by one of the layers of the ionosphere, and then returns directly to the receiver. In between is the **skip distance**, in which the signal path is not near enough to Earth for it to be received at all. However, the situation may be more complicated than this because, as we have seen, both land and (particularly) sea are themselves effective reflectors for radio waves. In the HF band, where the wavelength is between 10 and 100 metres, for the most part the reflecting surface is relatively smooth. Near-specular reflection from the ground is therefore commonplace, making possible multiple-hop transmission. In the two-hop case, the transmission is reflected from the ionosphere back to ground, reflected there back up to the ionosphere again and thence down once more to the receiver. Three and even four hops are possible but rare.

Designing for these multiple-hop paths is complex. The transmission frequency must, of course, be safely below the muf at both ionosphere reflection locations, and they may not be the same, since one might be during the day when the other is at night. Sometimes either one- or two-hop solutions to the propagation requirement can be found, using different angles of elevation of the antenna. It is impossible to say which will be best without knowing the mufs for the reflection locations.

Given the angle of launch, and supposing the critical frequencies are also known (not locally to the transmitter but for the parts of the ionosphere where the reflection is expected to take place), the mufs can be calculated for both single and multiple-hop cases. For a typical noon critical frequency of 10 MHz this will range from 10.8 MHz at an antenna elevation of 70°, up to 57 MHz for an elevation of 10°. All higher frequency transmissions will escape into space. At night these values may fall by half. Evidently, ionospheric propagation is only possible up to the HF band, and sometimes the lower fringes of the VHF under favourable conditions.

Because the gain of a transmitting antenna is proportional to the square of frequency in the usual case of constant physical size, it is desirable to work as near as possible to the muf. In practice the prediction of the muf is not sufficiently reliable to allow safe working much above 85% of its value. Intercontinental transmissions may be during the day over some parts of the path and at night over others; it is the lowest muf over the path which sets the limit on usable frequency. A **least usable frequency (luf)** can also be defined, which depends on antenna and receiver parameters also transmitter power, and is the least frequency at which an acceptable S/N ratio is obtained at the receiver.

3.5 Ionospheric propagation in practice

The D layer, at a height below 80 km, is characterized by a very high rate of recombination due to the relatively high air pressure, so it exists only in daytime and vanishes quickly at dusk. For this reason the electron density is low, as therefore is its critical frequency, so it is penetrated by all radio transmissions above the MF band. When the D region is present, radio energy of long enough wavelength to interact with the electrons present is quickly dissipated via collisions by electrons with the many surrounding air molecules. This is what happens to MF signals. They are rapidly attenuated by the D layer and ionospheric reflection is virtually non-existent; in this band only the surface wave can be received strongly during daytime. It is therefore around surface wave propagation that the service is designed.

At night (particularly in winter) the D layer becomes vestigial due to recombination, and the electron concentration required for MF reflection is now found about 50 km higher, in the E layer, where the atmosphere is so thin that much less absorption takes place. As a consequence many sky-wave transmissions (ionospherically propagated via the E layer) come in from distant transmitters, filling the band with signals and greatly increasing the probability of interference between transmissions. The winter nightly cacophony on the MF (medium wave) broadcast band is a result of this unwanted sky-wave propagation, and is evident to anybody who cares to turn on a broadcast receiver. To the broadcasting engineer MF ionospheric propagation is wholly negative, spoiling night-time winter service. So far as LF frequencies are concerned, even at its weakest the D layer has sufficient electrons to trap them, and there is no sky-wave transmission at all.

By contrast, at higher frequencies, where surface wave propagation is of much less importance (except to the military and for rural radiotelephone), ionospheric propagation is exploited positively. Intercontinental transmissions are almost always possible at some frequency in the HF band (3–30 MHz). Radio quanta in this energy range are reflected by the E or F regions and also by the Earth's surface; so that multiple hops are not uncommon, connecting points on opposite sides of the globe. Still shorter waves (higher VHF, UHF and above) will be beyond the critical frequency even of the F2 layer, and will consequently penetrate the ionosphere and escape into space.

However, as might be expected considering the physical mechanisms by which it is produced, the ionosphere is a far from stable propagation medium, even at HF. There are many causes of variation, some regular and others irregular. As already noted, in addition to the diurnal variation, ionization in winter is less than in summer, and there is also a longer-term variation due to cyclic change in the Sun's activity, usually called the sunspot cycle because it corresponds with the variation in the number of observable dark 'spots' on the Sun's disc. This cycle has an eleven year period.

Multipath propagation occurs when there exist two or more different viable propagation routes, for example surface wave and sky-wave (particularly at MF) or single and multihop sky-wave paths (at HF). As always it greatly modifies the characteristics of the received signal. We begin by considering MF with sky wave and surface wave interfering over an idealized flat earth. The case is exactly as for terrestrial multipath, except that the height h is no longer small compared with the distance $2d$.

The sky-wave path length minus surface wave path length difference is:

$$\Delta = 2\left(\frac{h}{\sin\theta} - d\right) \tag{3.13}$$

so the phase difference is:

$$\psi = \frac{4\pi f}{c}\left(\frac{h}{\sin\theta} - d\right) \tag{3.14}$$

The resultant of the two wave functions will pass through a zero when this phase difference is a multiple of π and maxima at multiples of 2π. Also, since h is not constant, approximately:

$$\frac{d\psi}{dt} = \frac{4\pi f \cdot v}{c \cdot \sin\theta} \quad \text{where } v = \frac{dh}{dt} \tag{3.15}$$

Thus the phase of the two waves varies continuously if the ionosphere moves up (or down) with a continuous velocity, such as in the evening when recombination is causing it to move higher.

The received signal passes through successive maxima and minima in a cyclic manner with a period T approximated by:

$$T = \frac{c}{2vf} \cdot \sin\theta \tag{3.16}$$

Maxima occur when the phase difference is an even multiple of π and minima when an odd multiple; as the height of the ionospheric layer concerned changes the signal strength cycles through them, which is called **fading**. However, because frequency appears in these expressions, at any moment of time fading is different at different frequencies. This effect is known as **selective fading**. Note that when the path difference is short enough, selective fading will not be significant over the particular bandwidth of the transmissions in use. The term **flat fading** is then used to make this explicit.

Suppose that n is the number of wavelengths in the path difference at a frequency f_p where the received signal is at a maximum. The condition is:

$$n = \frac{2\pi f_p}{c} \cdot \left(\frac{h}{\sin \theta} - d \right) \tag{3.17}$$

At minima:

$$n + \frac{1}{2} = \frac{2\pi f_o}{c} \cdot \left(\frac{h}{\sin \theta} - d \right) \tag{3.18}$$

so:

$$\Delta f = f_p - f_o = \frac{f_p}{2n} = \frac{c}{2\Delta} \tag{3.19}$$

where Δf is the **coherence bandwidth**, and can be surprisingly small.

Selective fading spoils fringe reception of AM broadcasts in the MF band, particularly at night and in winter, due to sky-wave interference with the normal surface wave service. As a result the AM carrier will go through zeros at times when there is sideband energy present and this will cause severe distortion, because the receiver demodulator normally depends on the presence of a carrier. It disables the receiver AGC (which is carrier related) so the distorted signal is also loud.

So far the argument has been developed for the case of MF, where the multipath effect is between a surface and a sky-wave propagation, but the analysis of selective fading at HF follows closely similar lines, except that in this case it is two (or more) sky waves which are interfering, whether they arrive by the alternative one-hop and two-hop paths or even propagate around the world in opposite directions, for example east-about and west-about. Path lengths, and hence multipath differential delays, are an order of magnitude or two larger than in the MF case, and therefore n, the number of wavelengths in the path differential, is much larger. Despite the increased centre frequency, the result can be very low values for the coherence bandwidth. As a result HF transmissions are generally kept narrowband, to minimize the undesirable consequences of selective fading. SSB (see below) analogue speech transmission is in very widespread use, with an effective bandwidth of some 2.5 kHz. Data transmissions are generally restricted to low rate (formerly ≤ 2.4 kb/s, although this is progressively improving), and the design of modems takes into account selective fading.

Sky-wave propagation plays the major role in the intercontinental communications capability of the HF bands, using the F layers. It is true, however, that HF channels have the reputation of requiring skilful management to give long-distance communication, which even then is of poor transmission quality, principally due to selective fading. The reasons for these problems are intrinsic to the mode of propagation of radio waves via the ionosphere. As a result, in the 1960s and 1970s the use of HF declined somewhat, with interest turning to other intercontinental transmission technologies, particularly the use of satellites, despite the higher cost of these technologies. However, as a result of the rapidly falling cost of computing, 'intelligent' HF receiving equipment is now available. Self-optimizing systems either measure the characteristics of the ionosphere instantaneously by 'probe' transmissions, or obtain similar information by measuring the characteristics of known regular broadcast transmissions from distant sites. Improved signal processing and error correction (using ARQ) has resulted in better quality demodulated signals, and new adaptive modems have raised available data rates, to the point where digital speech transmission is economic. Together these developments have resulted in a resurgence of interest in the HF bands.

3.6 Satellite communications

The most important innovation in radio engineering of the second half of the twentieth century is communication with the aid of artificial satellites (Maral and Bousquet 1998). Arthur C. Clarke first proposed the use of an artificial satellite carrying a radio transponder in the October 1945 issue of *Wireless World*. A low-orbit satellite Sputnik 1 was launched by the former Soviet Union on 4 October 1957, and the United States sent its own satellite into orbit some three months later. In 1963 Syncom 2 was launched, the first synchronous satellite (its period matching the Earth's rotation). Since then more than 3000 satellites have been successfully established in orbit.

All such satellite systems must operate at radio frequencies high enough for the ionosphere to be penetrated, that is at VHF or above. Even so, particularly at lower frequencies, there is both refraction and absorption of the radio energy. This is worst when the satellite is near the horizon and propagation is at a very oblique angle to the ionospheric layers. Under these conditions at VHF and in the lower part of the UHF band there may be total loss of signal. By contrast, above about 500 MHz ionosphere effects almost vanish, so the lower frequencies are now scarcely used for satellites.

To a first approximation at least, the mechanics of artificial satellites are simple. The satellite must be placed in orbit around the Earth. This is achieved when it is given a velocity such that the gravitational attraction to Earth is balanced by forces generated by the curvature of its path combined with the velocity of its motion. The orbits are elliptical, although in many cases the ellipticity is so slight that they are very close to circular. As the altitude of the satellite increases the gravitational force lessens so its stable velocity decreases and the time it takes to circle the Earth (its period) increases. The gravitational force between the Earth (mass M) and a satellite of mass m distant r from the Earth's centre is:

$$F = \mu \frac{m}{r^2} \tag{3.20}$$

where $\mu = \dfrac{M}{\gamma}$ and γ is the gravitational constant.

At the same time, if it can be assumed that the orbit is circular and that ω is the angular velocity of the satellite in radians per second, then the centrifugal force is F^* where:

$$F^* = mr\omega^2$$

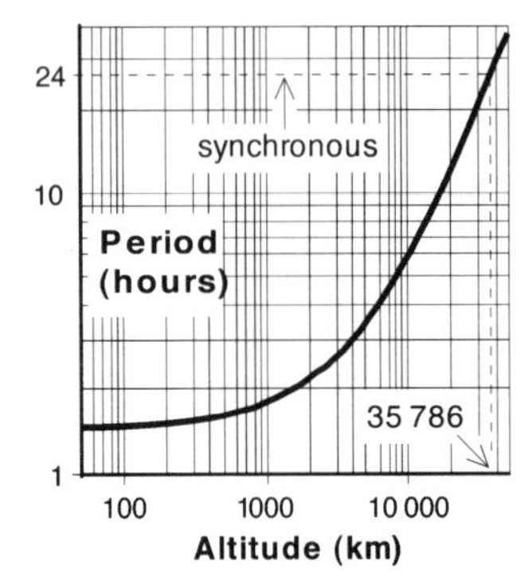

Fig. 3.5
Period versus altitude for an Earth satellite.

For stability in orbit $F = F^*$, also the period $T = 2\pi/\omega$ so:

$$T = \frac{2\pi r^{3/2}}{\mu^{1/2}} \tag{3.21}$$

But we can replace r by $(a + h)$, where a is the Earth's radius and h is the altitude of the satellite above the Earth's surface, so:

$$T = \frac{2\pi a^{3/2}}{\mu^{1/2}} \left(1 + \frac{h}{a}\right)^{3/2} \tag{3.22}$$

The constant μ is equal to 3.986×10^{14} MKS units and a is approximately 6378 km, so (Fig. 3.5):

$$T = 5.071 \times \left(1 + \frac{h}{6378 \times 10^6}\right)^{3/2} \times 10^3 \text{ seconds} \tag{3.23}$$

When the altitude is 35 786 km the period of the satellite is 24 hours, which is the **synchronous orbit**. If the satellite is positioned in an orbit directly over the equator it will move exactly in step with the part of the Earth's surface beneath it, so that to an observer on the ground it will appear to hang stationary in the sky. This is a **geostationary orbit**. By contrast, if the orbit is inclined to the equatorial plane there will still be some movement of the satellite as seen from Earth; a slight tilt of the orbital plane, for example, will result in an apparent north–south oscillatory movement.

3.7 Geostationary satellites (GEOs)

Because their capacity, in terms of the throughput of data, is determined solely by their on-board hardware, communications satellites have had a revolutionary impact on the practice of telecommunications. A measure of progress is the evolution from Intelsat I (1965) to Intelsat VI (1989). The earlier satellite weighed 68 kg and handled 480 telephone channels at a cost of $3.71 per channel hour, and had a working life of 1.5 years. By contrast, Intelsat VI weighed 3750 kg at launch and provides 80 000 channels, each at a cost of 4.4 cents per hour.

For use as a radio relay, a satellite in a geostationary orbit (a **GEO**) is particularly useful because ground-based antennas with a very narrow main lobe, and thus high gain, can be pointed at the satellite with little subsequent realignment. Similarly antennas on the satellite can be permanently aligned on fixed targets on Earth. This ability to use high gain antennas is essential because the principal disadvantage of geostationary satellites is their distance.

All GEOs receive the 'up' signal, increase its power, translate it to another frequency and re-radiate it back to Earth. Bearing in mind that the gain from the received signal to the transmitter output may be well over 100 dB, the frequency translation is essential because otherwise it would be impossible to prevent the powerful 'down' signals from leaking into the 'up' receiving antennas, causing unwanted feedback and system malfunction. Frequencies are allotted by international agreement at the periodic World Administrative Radio Conferences (WARC). So also are angular segments of the geostationary orbit, committed to particular national administrations and suitably located for the territories concerned.

A serious problem for geostationary satellites arises from orbital congestion. All GEOs must be in the same equatorial orbit, which therefore becomes very congested over the more heavily populated longitudes. If the same up-frequency were used for all satellites, the only way of directing signals from Earth towards one rather than another would be by exploiting the directivity of the ground station antenna. This has severe limitations, however, because of the distance from Earth to GEO. This is so large that every one degree of main-lobe width at the ground station corresponds to about a 600 km segment of orbit. It is hardly surprising that all the orbital locations serving Europe and the Americas were soon taken up. Since there are limits to the antenna directivity which can be engineered economically, the obvious solution to this problem is to operate different satellites at different frequencies (or sequencies, see below), so that they do not interfere with each other.

The EHF band is particularly attractive for satellite use, since EHF antennas can be highly directive without being too large. For a paraboloid or an array, for example, the gain and hence the directivity is proportional to the square of a characteristic dimension expressed in wavelengths, so for a fixed lobe width the dimensions of the antenna vary directly with the wavelength. Atmospheric absorption is not too harmful at EHF because of the high elevation angle of the ground station antenna, which means that the radio quanta quickly pass out of the atmosphere. The only real disadvantages of EHF in this service have long been that a watt of EHF power was considerably more expensive than a watt of SHF, and also EHF receivers were a few decibels less sensitive. Both of these problems are being progressively overcome. The use of EHF is therefore growing, and the problems of orbital congestion are being held off, for the time being.

We have already seen that many of the problems of using geostationary satellites arise from the relatively great distance from the Earth of the synchronous orbit. So how far away from an Earth station is the satellite? With increasing latitude the distance of an Earth station to the satellite increases from the equatorial value of 35 786 km to a polar 42 669 km. For some applications, like real-time speech, it is a disadvantage that the round trip to the satellite and back takes 0.24 seconds at the equator, rising to 0.27 seconds at the poles.

Inverse-square quanta spreading means that to a receiving antenna of one square metre aperture on the ground at the equator the path loss from the satellite is 162 dB, and the 'up' path loss is the same. Because the path loss is high, large EIRPs are required at the transmitters, but at the satellite the available transmitter power is limited to what solar panels can provide (70–100 watts per square metre). This implies the need for high gain, large aperture antennas on both the satellite and the ground station.

At the poles the path loss is greater by about 1.5 dB. However, at these latitudes other problems become severe, as the angle of elevation of the antenna gets less and less. The radio path becomes increasingly susceptible to near-surface losses and blocking by mountains or buildings, an effect known as **shadowing**. Although for fixed stations it is often possible to site the antenna to overcome the problem, it presents serious difficulties for land mobile installations, for example on vehicles used in city environments.

In Europe and many other areas of the world, high power geostationary **direct broadcast satellites (DBS)** are commonplace, broadcasting to individual domestic TV installations. An early example was the Astra DBS satellite, which had 16 TV channels, with a power per channel of 45 watts and a transmitting antenna gain of 35 dB. The EIRP in the central service area was +82 dBm (158.5 kW). Such high EIRP made possible the use of small receiving antennas, typically 0.6 m diameter in the principal service area and just a little larger in fringe areas. With the centre of its antenna 'footprint' located on the Franco-German border, the satellite gave satisfactory coverage over virtually the whole of the European Union, except for the extreme south of Italy. For services such as broadcasting, satellites are very economical of spectrum, because they use only a single transmission frequency to cover a very wide area.

3.8 Low orbit satellites

Geostationary satellites do some tasks very well indeed but not others, whether by reason of their large path loss, poor Arctic and

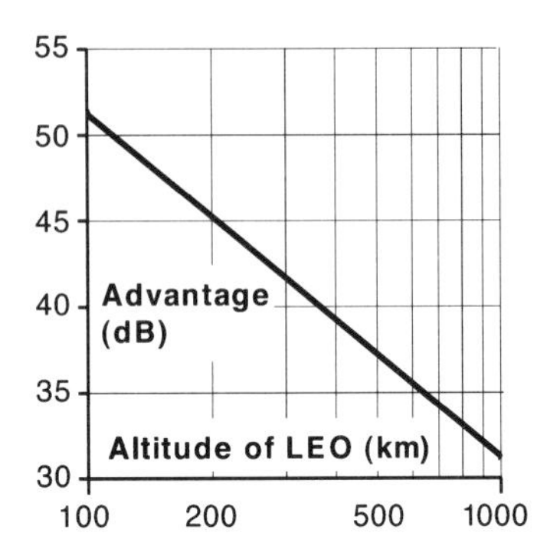

Fig. 3.6
Path loss advantage of an LEO over a GEO.

Antarctic coverage, or the long 'round trip' time delay. For this reason there is growing interest also in low Earth orbit satellites (LEOs). Often at only a few hundred kilometres altitude and frequently in polar orbits, they typically have periods of under two hours, and therefore any satellite will only be visible from a terrestrial site for a limited time, after which it will be obscured by the Earth until it 'rises' again on having completed an orbit. At any location they are only briefly (although frequently and predictably) receivable.

Low orbit satellites have a much more favourable power budget for radio links than GEOs, so terminal equipment could potentially be cheap (Fig. 3.6). They have two main disadvantages, however. The first is that the greater air resistance at low altitude gives rise to a shorter satellite life in orbit, which can be countered by repeated launchings. The second problem, at the ground station, is that the movement of the satellite across the sky means that either the ground antenna must track it or have a wide enough lobe that the satellite is receivable for an acceptable period of time at each transit.

Sophisticated low orbit systems have now been introduced that can show marked advantages against competing approaches. Iridium, launched commercially in 1998, is a system aimed initially at giving worldwide coverage to hand-held radiotelephones, but now offering

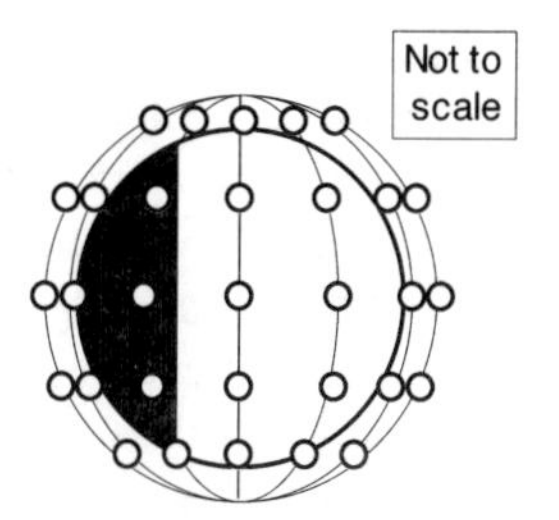

Fig. 3.7
The Iridium system of LEOs; 66 satellites in six polar orbits.

a growing range of other global services (Fig. 3.7). It was designed by Motorola Inc. and is managed by an international consortium. As many as 72 satellites are each in polar orbit at 780 km in six orbit planes of eleven functioning satellites plus one spare, thus providing continuous line-of-sight coverage to any place on earth. The expected life in orbit of the satellites, each weighing 689 kg, is 5 to 8 years. Satellite radio links to the ground are at 1.616–1.6265 GHz for digital voice communication to hand-held terminals. The symbol rate transmitted is 2.4 kilobaud and QPSK modulation is used. As well as digital speech, fax and data transmission are also supported. The main system ground stations communicate with the satellites at 29.1–29.3 GHz (up) and 19.4–19.6 GHz (down), whilst there are also 20 GHz links from each satellite to the one in front of it and the one behind in its particular orbit and to adjacent satellites in neighbouring orbits. Thus each satellite is in communication with four others. As satellites pass over them, service users on the ground 'hand off' from one satellite to another in a way exactly analogous to the 'hand-off' by mobile users in an ordinary cellular radiotelephone scheme. In fact Iridium has the same mathematics as a cellular system, the only difference being scale and the fact that it is the infrastructure which is moving.

Iridium was the first major and global commercial use of low orbit communications satellites. User terminals are already many and various, among them aircraft and ship installations, hand-held phones similar in size to cellular phones, solar-powered public call offices and a variety of data terminals. Between them they constitute a new worldwide and comprehensive telecommunications system, to complement and rival the established public network with which we are all familiar. Other ambitious LEO systems follow this lead. Giving universal coverage without the need for a terrestrial infrastructure, the likely significance of the LEO systems development is incalculable. Some argue that LEOs could provide all world telecommunications needs without a fixed network at all, but this view remains controversial at the present time. However, the commercial viability of these services has yet to be established.

See: *http://www.iridium.com/*

Questions

1. In Europe AM broadcasting stations are often received at night in winter over both surface and sky-wave types of propagation. Why does this not happen in summer daylight hours? If channels are spaced 9 kHz apart, what is the maximum path length difference in order that the coherence frequency shall not be less than the channel spacing? (17 km)

2. If, at a certain latitude, the average wait for a usable meteor-scatter channel is 2 minutes, what is the probability of having to wait 5 minutes? (just over 17%)

3. The coherence bandwidth of an HF transmission is 300 Hz due to both two-hop and one-hop transmission occurring simultaneously. What is the path difference between the two modes of propagation? (500 km) If the great-circle distance between the transmission and reception sites is 2000 km and the F2 layer is in use, what is the launch angle for the two modes of propagation? (about 23° for one hop, 42° for two) Could a Yagi antenna array be used to eliminate two-hop propagation? (No, but it could reduce it by several decibels)

Part Two

Optimizing Radio Spectrum Use

Chapter 4

Regulating the radio spectrum

Unregulated use of the radio spectrum, in which all were free to transmit whatever they liked, at any power level or in any format, would lead to chaos, high levels of interference and pollution, and a reduction in the utility of the radio spectrum for all users. Regulation makes it possible to minimize these problems, by ensuring the correct choice of operating frequencies and other technical factors in transmissions from a particular location and for a specific purpose, and by setting standards for the design of radio systems and of the equipment to be used both in transmission and reception.

4.1 The logic of regulation by international agreement

The mechanisms of propagation of the radio quanta in the different bands vary enormously, as we have seen. Each band has its own particular range of uses and demands its own distinctive pattern of regulation. This has prompted the development of a complex set of binding international agreements about the use of the electromagnetic spectrum (Withers 1991). These arrangements can seem complicated but their adoption is absolutely essential. Doing so by international agreement has important advantages:

1. Where radio propagation is possible beyond national borders (as it is in many bands) international agreements facilitate

successful international working while minimizing the risk of mutual interference.

2. For ships, aircraft and other international travel, use of radio while in transit is essential for traffic control and safety, as well as an important service to travellers, but is only made possible by international agreement on radio standards, spectrum use, and procedures.

3. By the greatest possible international standardization of the use of radio bands the manufacture of equipment can be undertaken on a worldwide basis, leading to a healthy world radio manufacturing industry, with all the innovations and price reduction that competition brings.

Regulating the use of the electromagnetic spectrum is the duty of the **International Telecommunications Union** (**ITU**), based in Geneva. Successor to the International Telegraph Union founded in 1865, this is the oldest of all intergovernmental bodies, and without it the world of telecommunications would be very different and much less effective. The ITU administers decisions on spectrum use which are collectively arrived at by all the member nations, their representatives meeting together at a **World Administrative Radio Conference (WARC)** held every few years. These are backed up from time to time by **Regional Administrative Radio Conferences**, as may prove necessary. At these meetings agreements are hammered out which are then binding on the national authorities.

See: *http://www.itu.int/ITU-R/index.html*

4.2 The national radio regulatory agencies

In accord with these international agreements, the detailed regulation of spectrum use is undertaken in each country by a national **regulatory authority**, backed by the force of national law. For the UK, the regulatory authority is the **Radiocommunications Agency** in London. All other developed nations have similar organizations;

for the US there is the **Federal Communications Commission** (**FCC**) in Washington DC.

See: *http://www.open.gov.uk/radiocom/rahome.htm*
also: *http://www.fcc.gov*

To explain the work undertaken by these agencies, some additional terminology will be useful. Frequency **allocation** consists of designating substantial blocks of spectrum for broad specific purposes, such as broadcasting, maritime, amateur, fixed links and so on. This is invariably settled by international agreement. Within allocations, **allotment** involves designating smaller blocks of frequencies to more specific user groups, such as public utilities and emergency services, the military, and cellular phone systems. Frequency **assignment** consists of designating a particular channel within an allotment for a particular user in a specified location. A nationally issued **licence** to use radio will often be based on a particular assignment.

As indicated, allocation is always and in all cases agreed on an international basis, as also is assignment in situations where the radio service concerned is transnational (in this case the stage of allotment is usually skipped). Shorter range radio systems, which remain largely within national boundaries, are allotted and assigned at the initiative of the national regulatory organizations, which may delegate assignment to other national organizations in some cases.

4.3 The technical challenge

How do the engineers who are advising the regulatory authorities make sense of these difficult problems? Their objective is to minimize spectrum occupancy, pollution and interference. Different types of radio use have very different characteristics, however. For example, the problems of spectrum pollution and interference for cellular mobile radio are quite different to those of terrestrial television broadcasting, and different again to those of satellite systems.

We begin by considering the minimization of spectrum occupancy and pollution associated with the transmitted signal. The analogue and digital cases must be treated separately. At the present time radio transmissions are commonplace in both of these forms; although digital techniques are steadily replacing analogue in many applications (and may replace them altogether one day) for the present both remain important.

Questions

1. Distinguish between frequency allocation, allotment and assignment. In what bands would you expect the last of these to be a purely national responsibility, and why?

2. What are the advantages of international agreement on spectrum regulation?

Chapter 5

Optimizing analogue radio transmissions

The designers of the earliest radio systems faced a problem: the moderate-quality audio waveforms of speech and music that they wished to transmit had significant frequency components between, say, 50 Hz and 5 kHz, but radio propagation was even then known only to be satisfactory at higher frequencies. They therefore needed to translate the audio frequencies up into a much higher band before transmission. They achieved this result by combining the audio signal with a suitable **carrier wave**, a sinusoid in the desired **radio frequency** (**RF**) range, much higher in frequency than the highest component of the signal waveform.

This process of combining a baseband signal with an RF carrier (in whatever way) is known as **modulation**. Note, in passing, that the carrier need not be sinusoidal, indeed the sinusoidal carrier must be regarded as a special case; systems using non-sinusoidal carriers are of increasing importance and will be reviewed subsequently. In the first radio transmitters, however, sinusoidal carriers were universal and modulation was by a process mathematically equivalent to simple multiplication.

5.1 Amplitude modulation

Consider a **baseband** waveform $e(t)$, which will be an **audio frequency** (**AF**) wave in the case of analogue speech or music or a **video**

frequency (**VF**) wave for television, and also a sinusoidal carrier wave $E \cdot \cos \omega_c t$. The modulated waveform is ideally $E_m(t)$ where:

$$E_m = E[1 + b \cdot e(t)] \cos(\omega_c t) \tag{5.1}$$

Without much loss of generality we may approximate the baseband waveform $e(t)$ as a Fourier series of the form:

$$e(t) = \sum c_r \cos(\omega_r t + \phi_r) \tag{5.2}$$

Hence:

$$E_m = E \cdot \left[1 + b \sum c_r \cos(\omega_r t + \phi_r)\right] \cdot \cos(\omega_c t) \tag{5.3}$$

One or two points are immediately obvious. In equation (5.3), the term in square brackets is the time-dependent amplitude of a sine wave at the carrier frequency. Since the cosines in the sigma term cannot fall below -1, the amplitude will always be positive provided that:

$$1 - b \sum c_r \geq 0 \tag{5.4}$$

In this case the modulated waveform is simply a well-behaved sinusoid of constant frequency which varies in amplitude between A_{max} and A_{min} where:

$$A_{max} = E \cdot \left[1 + b \sum c_r\right] \quad \text{and} \quad A_{min} = E \cdot \left[1 - b \sum c_r\right]$$

This type of modulation is known as **amplitude modulation** (**AM**) or sometimes as **full-carrier AM** (for reasons which will become apparent). The **modulation index**, usually expressed as a percentage, is defined by:

$$\left.\begin{aligned} m &= \frac{A_{max} - A_{min}}{A_{max} + A_{min}} \cdot 100\% \\ &= b \sum c_r \cdot 100\% \end{aligned}\right\} \tag{5.5}$$

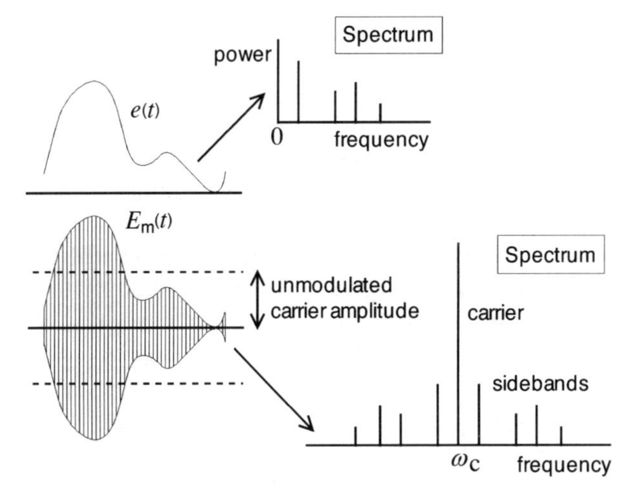

Fig. 5.1
A full AM waveform compared with its baseband modulating signal.

When the modulation index is 100%, the envelope of the modulated waveform just falls to zero at modulation minima while at maxima it reaches double the unmodulated value (Fig. 5.1). Thus the peak power is four times (+6 dB) relative to the unmodulated power.

So much for the AM wave in the time domain, but what does it look like in the frequency domain? What is its spectrum? The answer presents no difficulty. In equation (5.3) we have the sum of the products of pairs of cosines. This expression can be expanded by using the well-known formula for the product of cosines:

$$\cos\alpha\cdot\cos\beta = \tfrac{1}{2}[\cos(\alpha+\beta) + \cos(\alpha-\beta)]$$

Applied to each pair in the summation this leads to:

$$E_{\mathrm{m}} = E\cdot\left[\cos(\omega_{\mathrm{c}}t) + b\sum\frac{c_{\mathrm{r}}}{2}\left\{\begin{matrix}\cos([\omega_{\mathrm{c}}-\omega_{\mathrm{r}}]t-\phi_{\mathrm{r}}) \\ +\cos([\omega_{\mathrm{c}}+\omega_{\mathrm{r}}]t+\phi_{\mathrm{r}})\end{matrix}\right\}\right] \tag{5.6}$$

Evidently, in the frequency domain the modulated waveform contains three groups of components. The first term on the RHS of equation (5.6) is at the carrier frequency and has constant amplitude. The second term consists of a series of spectrum lines, the **lower sidebands**, at frequencies displaced below the carrier frequency by an amount equal to the frequency of the modulating component from which they derive. The third term consists of a similar series of **upper sidebands**, displaced from the carrier frequency in the same way but now above rather than below.

When this waveform is received it is **demodulated** to extract $e(t)$. Supposing that the carrier frequency is known at the point of demodulation, consider the effect of multiplying the upper sidebands only by a sinusoid at the carrier frequency.

$$
\begin{aligned}
&b\sum\frac{c_{\mathrm{r}}}{2}\cos([\omega_{\mathrm{c}}+\omega_{\mathrm{r}}]t+\phi_{\mathrm{r}})\cdot\cos\omega_{\mathrm{c}}t \\
&\quad=b\sum\frac{c_{\mathrm{r}}}{4}\cos(\omega_{\mathrm{r}}t+\phi_{\mathrm{r}})+b\sum\frac{c_{\mathrm{r}}}{4}\cos([2\omega+\omega_{\mathrm{r}}]t+\phi_{\mathrm{r}}) \\
&\quad=\frac{b}{4}\cdot e(t)+b\sum\frac{c_{\mathrm{r}}}{4}\cos([2\omega_{\mathrm{c}}+\omega_{\mathrm{r}}]t+\phi_{\mathrm{r}}) \qquad (5.7)
\end{aligned}
$$

The first term of equation (5.7) is proportional to the wanted modulating signal, whilst the second is a group of Fourier components at a very much higher frequency, easily removed by low-pass filtering. This is sufficient to demonstrate that the set of upper sidebands alone evidently contains all the information required to derive $e(t)$, and it is obvious that the same is true of the set of lower sidebands. It is not difficult to see that the modulated waveform of equations (5.3) and (5.6) contains redundant elements so far as the potential for demodulation is concerned, namely the carrier component and one complete set of sidebands.

5.2 Why redundant the components in AM?

All unnecessary radiated energy constitutes spectrum pollution, so why are they present, these redundant carrier and sideband components? There are two reasons. The first is that when AM was invented the electronic hardware required to do more than simple (and approximate) multiplication was not available, and AM became well established before this situation changed, so to some extent they are a historical survival.

The second, more powerful reason is that the carrier component simplified the process of demodulation in early receivers, since with the carrier and both sets of sidebands present the envelope of the modulated signal corresponds to $e(t)$. Only a simple rectifier and low-pass filter are needed to extract the wanted signal.

However, the penalty for using this simple form of AM is high. Since only one set of sidebands is actually needed to communicate the wanted signal (whether the upper or lower set) it follows that technically the carrier component and the other set of sidebands constitute spectrum pollution. They are not necessary for the communication process, yet can cause degradation of service for other users. This is by no means a merely theoretical point, particularly as the proportion of the power in the unwanted components is so large.

Taking the simple case of sinusoidal modulation, so that:

$$e(t) = c \cdot \cos(\omega_m t)$$

Assuming 100% modulation (so $b \cdot c = 1$) equation (5.6) simplifies to:

$$E_m = E \cdot [\cos(\omega_c t) + \tfrac{1}{2}\cos([\omega_c - \omega_m]t) + \tfrac{1}{2}\cos([\omega_c + \omega_m]t)] \quad (5.7)$$

This spectrum consists of a carrier, one upper sideband and one lower sideband. The power of each sideband is one-quarter of that of the carrier component, or to put it another way, relative to the one sideband needed to communicate the baseband signal, the carrier component is at +6 dB and the total power +8 dB (×6). The excess power constitutes serious spectrum pollution, yet the case of sinusoidal modulation considered so far is actually more favourable than many real situations.

Where the transmitted waveform has a higher peak-to-mean ratio than a sine wave things get rapidly worse. The modulation index cannot be allowed to exceed 100% on peaks, so for much of the time, when the signal is well below its peak value, the modulation index will be considerably lower. For example, natural analogue speech has such a large peak-to-mean ratio that it is impossible to sustain more than an average 10% modulation during an utterance, and even considerable automatic speech compression (making the quiet sounds appear louder and the loud softer using a **compander**) does not yield better than 30%. In the frequent silences between utterances the modulation index necessarily falls to zero, reducing the medium-term average still further. Music can be even worse than speech if a good dynamic range is to be retained. Even with compression, well over 90% of the energy radiated by radio systems carrying analogue speech or music is carrier. Yet all this pollutant energy has no higher purpose than to simplify the demodulator in the receiver.

So far as the 'unwanted' set of sidebands is concerned, an additional undesirable effect is to double the spectrum width of the radio signal, occupying an adjacent channel which would otherwise be available to another user. As an electromagnetic pollutant the unwanted sidebands can by no means be ignored either. Although always substantially less than the carrier, their power is nevertheless exactly the same as that of the wanted sidebands.

5.3 Pollution by AM signals

Transmission impairments arise when two signals are received in the same nominal channel, only one of which carries the wanted modulation. When both signals are AM, the consequences of the interference can be assessed for all the spectrum components of the signal, but since, as we have seen, the carriers are the largest component it is the interaction between these which is likely to predominate.

Carrier interactions are particularly simple, since each carrier is a constant amplitude sinusoid. In the very unlikely event that they are both at exactly the same frequency, their effect is just dependent on their phase relationship. If they were in-phase their amplitudes would just add, producing an enhanced total carrier and so a reduction of the apparent modulation index of the wanted signal. Depending on the receiver design, the result may be no worse than some loss of demodulated baseband signal amplitude. By contrast, if they were in antiphase the total carrier would be reduced, which might result in defective demodulation with serious signal distortion from a seemingly over-100% modulated signal. Usually though there will be a frequency difference between the two carriers $\Delta\omega$ and writing E_r for the resulting carrier:

$$E_r = E_w \cos(\omega t) + E_i \cos([\omega + \Delta\omega]t)$$

where E_w is the amplitude of the wanted carrier and E_i is that of the interferer, assumed smaller. Taking the sums of cosines in the usual way:

$$E_r = (E_w - E_i)\cos(\omega t) + 2E_i \cos\left(\left[\omega + \frac{\Delta\omega}{2}\right]t\right) \cdot \cos\left(\frac{\Delta\omega}{2}t\right) \tag{5.8}$$

The first term on the RHS is merely a carrier component, but the second is more interesting. The envelope of this wave looks like the modulus of a cosine of frequency $\Delta\omega/2$, or to put it another way, it looks like a full-wave rectified version of that cosine function. Its spectrum will show energy at $\Delta\omega$ and all its even harmonics. At the output of the AM demodulator (depending on just how it works) a component due to carrier interaction will be present at $\Delta\omega$ (the carrier **heterodyne** or '**beat**') and possibly at its harmonics, although this depends on whether these fall above or below the cut-off of the low-pass filter used in the demodulator.

In any speech or music radio system the carrier beats are heard as continual whistles or lower notes, whilst in a television image a pattern of semi-transparent bars is seen, usually at an angle to the vertical and often drifting in both position and angle as the phase and frequency relationships between wanted signal and interference vary with time. Both for sound and vision the effects of carrier interactions are seriously disturbing to the listener or viewer. By contrast sideband interactions are more transient, and although unpleasant if severe, can exist at a significant level before they cause any marked negative subjective response. For analogue AM transmissions it is therefore the carrier interactions which are the dominant source of interference.

5.4 Frequency modulation

Although all the early transmitters were amplitude modulated, from the late 1930s **frequency modulation (FM)** became increasingly fashionable. From a spectrum conservation point of view this was something of a disaster.

In the case of FM with, once again, a baseband waveform $e(t)$ and a sinusoidal carrier $E \cdot \cos \omega_c t$, the modulated waveform is ideally $E_{fm}(t)$ where:

$$E_{fm} = E \cos(\omega_c[1 + b \cdot e(t)]t) \tag{5.9}$$

Now it is the amplitude of the wave which is constant but the frequency varies in accordance with the modulation. Using once again the expansion of $e(t)$ as a Fourier series

$$e(t) = \sum c_r \cdot \cos(\omega_r t + \phi_r)$$

Evidently then:

$$E_{fm} = E \cos\left(\omega_c\left[1 + b \sum c_r \cos(\omega_r t + \phi_r)\right] t\right)$$

The maximum **frequency deviation** from the central value is:

$$\Delta\omega = \pm\omega_c \cdot b \sum c_r \tag{5.10}$$

If the maximum value of ω_r is ω_{max}, the ratio $\Delta\omega/\omega_{max}$ is called the **deviation ratio**. Sometimes the ratio $\Delta\omega/\omega_c$ is called the modulation index, by analogy with AM (see equation (5.5)), but this term has also been used in other senses for FM, and is best avoided.

Obtaining the spectrum of an FM transmission of this kind is much more difficult than for AM because of the terms in t^2 in the cosines.

Once again the outcome is (in general) a component at carrier frequency and also upper and lower sidebands corresponding to each modulating frequency, but now in addition there are upper and lower sidebands for the harmonics of every modulating frequency, so that in theory the spectrum extends from zero frequency to infinity. The amplitudes of the various components can be shown to correspond to Bessel functions depending on the deviation ratio, and in practice for all real cases these rapidly fall to very low values at frequencies remote from the carrier. Thus although the spectrum is notionally infinite, it can be constrained to a finite bandwidth without significant distortion. As a rule of thumb, satisfactory performance can be expected if the spectrum is allowed to extend over a total bandwidth of at least $2(\Delta\omega + \omega_{max})$ centred on the carrier.

Since this is always more than the $2\omega_{max}$ occupied by an AM transmission, it might be wondered why there has ever been any enthusiasm for FM. Three reasons are significant:

1. In the early days of FM it was thought that substantially better protection against noise and interference could be expected, and that this would have system advantages. The theoretical basis of the argument was the assumption of free-space propagation, sinusoidal modulation and constant signal levels. In the event, with speech or music as modulation, the hoped-for advantages were not realized. Reception by scattering propagation made matters still worse.

2. At the time the need for spectrum conservation was not understood, since at that early stage in the development of radio there was little spectrum congestion.

3. Finally, the most important reason for the use of FM was that it led to simple, low-cost transmitters and receivers, since they were not required to respond linearly to amplitude variations. So FM, though technically disappointing, was a 'cheap and cheerful' system that eased early exploitation of the VHF and UHF bands when hardware was less sophisticated and far more costly (in real terms) than today.

What of the interference and pollution prospects for analogue FM transmissions? As with AM, the answer is dominated by the fact that analogue transmissions, whether of music or speech, are severely undermodulated most of the time. So although it is true that for certain deviation ratios with FM the carrier disappears altogether, in practice the average reduction of carrier is small and transmitters spend a good deal of their time unmodulated, radiating nothing but carrier. The large carrier component is, once again, the major source of impairment for FM as for AM. The audible effect is a little different in the two cases, with more audible distortion in the FM case, although heterodyne beats and whistles are still a major feature.

5.5 Reducing spectrum pollution in analogue systems

From as early as the 1930s attempts were being made to improve on simple AM and FM, although not at first from the standpoint of spectrum conservation. The waste of transmitter power in radiating a large carrier component seemed, at the time, more important than the pollution the radiated carrier caused, particularly in the case of high powered transmissions such as television broadcasting. There was also an awareness of the bandwidth disadvantage of radiating two sets of sidebands, where only one is strictly necessary, making for less effective receiver designs. What would be the most spectrum-conserving approach to radio transmission, in the simple case of a single transmitter?

We look first at the baseband signal, ensuring that there is no redundancy there. In the case of digital signals the baseband offers many possibilities for greater efficiency, largely through data compression. We shall look at this later. In the case of analogue speech and music there is less to be done, although commonly the frequency range is limited by suitable audio filters. Thus when good music quality is required the range may be 50 Hz–15 kHz, but for lower quality sound broadcasting the upper cut-off may be only 4 kHz while speech communication circuits usually accept AF limited to 300 Hz–3 kHz or less. Analogue colour television baseband signals are considerably processed to contain their bandwidth, exploiting the knowledge that some impairments of the signal do not result in noticeable degradation of the picture.

Suppressing the carrier of an AM transmission is relatively easy; if the multiplier is modified from that in equation (5.1) so that the modulated signal is now:

$$E_{\mathrm{dsbsc}} = e(t)\cdot E\cos(\omega_{\mathrm{c}}t) \tag{5.11}$$

this is equivalent to:

$$E_{\mathrm{dsbsc}} = E\cdot\left[\sum\frac{c_{\mathrm{r}}}{2}\{\cos([\omega_{\mathrm{c}} - \omega_{\mathrm{r}}]t - \phi_{\mathrm{r}}) + \cos([\omega_{\mathrm{c}} + \omega_{\mathrm{r}}]t + \phi_{\mathrm{r}})\}\right] \tag{5.12}$$

In the spectrum of this waveform both sets of sidebands are present but there is no carrier component. In consequence this is known as a **double-sideband suppressed carrier (DSBSC)** signal. Although it still occupies the same bandwidth as an AM signal, at least the carrier component is not transmitted, so both the pollution and waste of transmitter power from this cause are overcome (Fig. 5.2). This has led to occasional proposals for its use in practical radio systems, but there is a problem when it comes to demodulation.

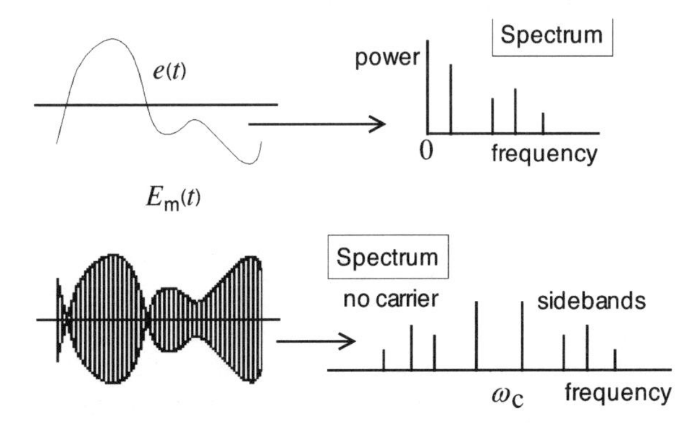

Fig. 5.2
A DSBSC waveform compared with its baseband modulating signal.

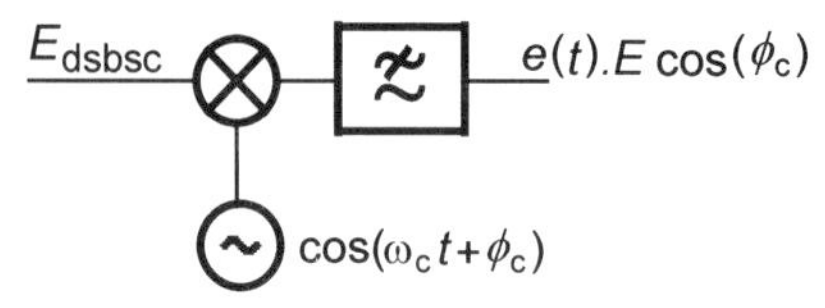

Fig. 5.3
Demodulating a DSBSC signal.

The envelope of the waveform of equation (5.11) certainly does not correspond to the baseband waveform (as it does with full-carrier AM), indeed a little thought will show that the envelope actually follows the modulus of $e(t)$. Equation (5.11) could perfectly well be written as:

$$E_{\mathrm{dsbsc}} = |e(t)| \cdot E\cos(\omega_c t + \alpha) \tag{5.13}$$

where $\alpha = 0$ if $e(t)$ + ve, $\alpha = \pi$ if $e(t)$ – ve.

In engineering terms, the envelope is a full-wave rectified version of the modulation and a half-cycle phase discontinuity in the RF occurs wherever the modulation passes through zero. This means that a simple rectifier and low-pass filter cannot be used as the demodulator, as it could with full-carrier AM. Something more sophisticated is necessary.

Suppose that the waveform of equation (5.12) is multiplied by a carrier component and a low-pass filtering function F (Fig. 5.3) then the demodulated waveform is given by:

$$E_d = F(e_y)$$

$$e_y = E \cdot \left[\sum \frac{c_r}{2} \cdot \{\cos([\omega_c - \omega_r]t - \phi_r)\} + \{\cos([\omega_c + \omega_r]t + \phi_r)\} \right] \cdot \cos(\omega_c t + \phi_c) \tag{5.14}$$

then it follows that:

$$E_d = E \cdot \left[\sum \frac{c_r}{2} \cdot \{\cos(\omega_r t + \phi_r + \phi_c) + \cos(\omega_r t + \phi_r - \phi_c)\} \right]$$

or:

$$E_{\mathrm{d}} = E \cdot \left[\sum c_{\mathrm{r}} \cdot \cos(\omega_{\mathrm{r}} t + \phi_{\mathrm{r}}) \right] \cdot \cos(\phi_{\mathrm{c}}) = e(t) \cdot E \cos(\phi_{\mathrm{c}}) \quad (5.15)$$

The baseband waveform has been recovered, but now multiplied by the cosine of the carrier phase angle, which must clearly be held to zero for maximum output. Thus a locally generated carrier may be used to demodulate the DSBSC waveform (by multiplication) but it must be not only of the right frequency but also be locked in phase (**coherent**) with the notional carrier of the transmitted signal. No carrier component is available for phase-locking, however, and although theoretically the required phase information could be derived from the sidebands (for example, using the so-called Costas loop demodulator) for discontinuous waveforms (like speech or music) this proves difficult to implement. In consequence DSBSC is not used in analogue systems.

It would be possible to add back a small carrier component, too low in amplitude to represent either a significant waste of power or a serious pollution threat, and to use this to synchronize the demodulator. This type of **double sideband diminished carrier (DSBDC)** transmission can achieve most of the advantages of DSBSC with easier implementation of the demodulator. Typically the carrier component may be some 10–20 dB below its full-AM value, reducing interactions by 20–40 dB if all transmissions are the same.

Questions

1. If a full AM transmitter is 30% modulated (typical for companded speech), what proportion of the average radiated power is carrier? (96%)

2. Why was FM widely used in the past, despite the waste of spectrum it entails?

3. A full AM transmitter is replaced by a DSBDC transmitter of the same peak power with pilot power 20 dB below peak. By what factor will the S/N ratio of the demodulated signal improve? Assume a rectifier demodulator in the case of AM. (+6 dB)

CHAPTER 6

TOWARD THE THEORETICAL IDEAL

Although the elimination of the AM carrier produces a valuable reduction in potentially polluting RF radiated power, it is not a complete answer to the problem of optimal signal form even in the analogue case. Demodulation of DSBSC is difficult and obviously the radiation of two sets of sidebands is still redundant. DSBDC delivers almost all the radiated power savings of DSBSC and is easier to demodulate but still has excessive bandwidth.

So, given a baseband analogue signal in minimal form, how is it optimally modulated into the desired RF band? Consider, once again, a baseband waveform:

$$e(t) = \sum_{\text{all } r} c_{\text{r}} \cdot \cos(\omega_{\text{r}} t + \phi_{\text{r}})$$

For minimum RF spectrum occupancy, this signal should be converted into a form where one spectrum line at RF coincides with each component of the baseband signal, so that the modulated signal is:

$$E_{\text{lm}} = \sum C_{\text{r}} \cdot \cos([\omega_{\text{c}} + \omega_{\text{r}}]t + \phi_{\text{r}}) \tag{6.1}$$

In this case each component of the baseband signal has been linearly translated up to a higher frequency, its frequency shifted by an amount equal to the carrier frequency (Fig. 6.1). There has also been a linear scaling of the amplitude of each component to give the required transmitted power. This type of modulation,

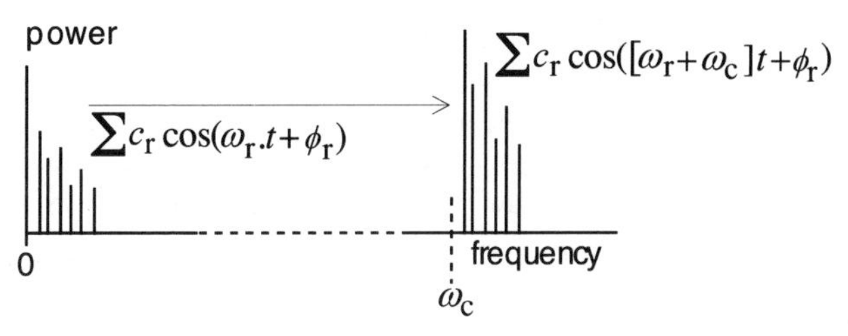

Fig. 6.1
Linear modulation (LM) involves the upward linear translation in frequency and power of the baseband spectrum without changing it at all in other respects.

which represents the theoretical optimum, is known as **linear modulation (LM)**, and specifically the **sum** form. An alternative, and equally spectrum conserving, variant is the **difference** form:

$$E_{lm} = \sum C_r \cdot \cos([\omega_c - \omega_r]t + \phi_r) \tag{6.2}$$

Note the similarity of both forms of linear modulation to the sidebands generated in AM, that in equation (6.1) to the upper set of sidebands and equation (6.2) to the lower. For this reason a variant of linear modulation was referred to as **single sideband** modulation **(SSB)**. However, this name is historically associated with a particular and limited realization of the LM idea, to be described subsequently, and its general use for linear modulation systems would be inappropriate. Note also that LM is not restricted to use with analogue signals, indeed its primary importance is with digital transmissions, but more of that later.

6.1 LM: the first method

Although the potential advantages of linear modulation were recognized at a very early stage in the development of radio technology, its use was inhibited by the difficulty (with early, undeveloped technology) in both generating and demodulating LM signals. Today we might consider using digital signal processing to perform a fast Fourier transform on the baseband signal, add a constant (equal to the carrier frequency) to each frequency term, and then perform an inverse transform to arrive at the desired final waveform. However, technology of this sophistication demands much computing power, and only became available in the last few years of the twentieth century, so early pioneers in the use of LM had to adopt a less sophisticated approach.

An obvious method is to subject the DSBSC waveform to further processing so as to eliminate one of the sets of sidebands, the outcome of which is an LM signal. Considering the expression for DSBSC in equation (5.12), and supposing that the Fourier components of the baseband waveform range from a minimum frequency of ω_{min} to a maximum frequency of ω_{max} then the lower ensemble of sidebands (corresponding to the difference LM waveform) is in the frequency range from $(\omega_c - \omega_{max})$ to $(\omega_c - \omega_{min})$ and the upper ensemble (the sum LM waveform) lies between $(\omega_c + \omega_{min})$ and $(\omega_c + \omega_{max})$.

Using a high-pass filter that passes all above $(\omega_c + \omega_{min})$ and stops everything below $(\omega_c - \omega_{min})$ it would be possible to select out the sum LM signal only (Fig. 6.2). Similarly a low-pass filter could select the difference LM signal. This was the earliest-used method of generating an LM (or SSB) signal, but it has problems, in that the specification of the filter can be very hard to meet. The cut-off of the filter must be such that its attenuation (in either case) goes from minimum to near maximum in a frequency difference of only $2\omega_{min}$. Since the minimum frequency to be reproduced might be (for music) as low as 50 Hz, this would mean the filter making its transition from low to high attenuation in only 100 Hz, very difficult

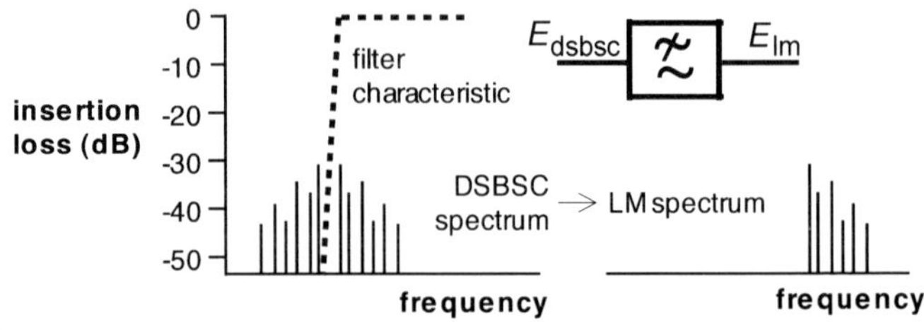

Fig. 6.2
The first (filter) method of generating linear modulation (LM).

if the carrier frequency is a few hundred kilohertz or higher. Such a specification could be met using modern digital filtering techniques, but in an earlier period when only LC or electromechanical filters were available it presented great difficulties. A lower carrier frequency would help, but this would require a subsequent additional frequency conversion to place the signal in the desired (higher) radio band, or possibly even two conversions. Transmitter designers found that this degree of complexity was not easy to engineer. So the filter method – sometimes named the **first method** – was used primarily with low quality voice circuits, where the minimum transmitted baseband frequency could be 300 Hz (or perhaps even higher) giving the filter at least 600 Hz in which to make its transition.

Two other methods of generating the LM signal were used in substitution, and since they form the basis of many present-day digital algorithms for generating LM signals, they will be described briefly.

6.2 LM: the second method

The **phasing** or **second method** relies on phase cancellation of the undesired sidebands (Fig. 6.3). Two DSBSC signals are generated, using two carriers at the same frequency, but one phase-shifted

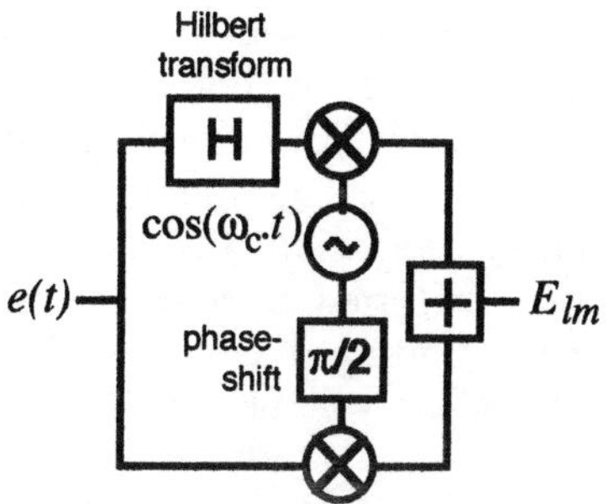

Fig. 6.3
The second (phasing) method of generating LM.

relative to the other by $\pi/2$ radians (90°). The result of the first of these, assuming a phase-shifted carrier, is:

$$E_1 = E \cdot \left[\sum \frac{c_r}{2} \left\{ \cos\left([\omega_c - \omega_r]t - \phi_r + \frac{\pi}{2} \right) \right. \right.$$
$$\left. \left. + \cos\left([\omega_c + \omega_r]t + \phi_r + \frac{\pi}{2} \right) \right\} \right] \tag{6.3}$$

The second DSBSC signal is generated with a version of the baseband signal $e(t)$ which has been passed through a network which implements a **Hilbert transform**, producing a new waveform:

$$e_H(t) = H[e(t)] = \sum c_r \cos\left(\omega_r t + \phi_r + \frac{\pi}{2} \right) \tag{6.4}$$

The Hilbert transform shifts the phase of every Fourier component by $\pi/2$ radians. This operation can be approximated over a limited frequency range by an LC network, or can be performed more accurately using digital signal processing. If the transformed baseband waveform is now multiplied by the carrier without phase shift the result is:

$$E_2 = E \cdot \left[\sum \frac{c_r}{2} \left\{ \cos\left([\omega_c - \omega_r]t - \phi_r - \frac{\pi}{2} \right) \right. \right.$$
$$\left. \left. + \cos\left([\omega_c + \omega_r]t + \phi_r + \frac{\pi}{2} \right) \right\} \right] \tag{6.5}$$

Comparing equations (6.3) and (6.5), it will be seen that both represent DSBSC waveforms, but the lower sidebands of E_1 are phase advanced by $\pi/2$ whilst those of E_2 are phase retarded by the same amount. The lower sidebands of the two waveforms are therefore in phase opposition. Clearly $(E_1 + E_2)$ will be an LM waveform of the sum form (upper sidebands only) whilst $(E_1 - E_2)$ will be an LM waveform of the difference form (lower sidebands); either can therefore easily be obtained.

The phasing method of LM generation seems neatly to side-step the filter problem, but it relies on the cancellation of two relatively large signals (the unwanted sideband ensembles) so that quite small errors in the processing lead to incomplete cancellation and significant residual unwanted sidebands. If the transfer characteristics of the two channels were to differ by only 1% in amplitude (easily possible using analogue Hilbert transform networks), the suppression of the unwanted sidebands would be only by 40 dB, which will not be enough if it is desired to receive relatively weak signals in the adjacent radio channel. However, with digital signal processing better than 60 dB suppression is often achieved.

6.3 LM: the third method

Because the Hilbert transform is not easily realized, the **third** or **Weaver method** has sometimes been preferred (Fig. 6.4). In this variant, the baseband signal is multiplied (in two parallel channels) by a sinusoidal function located in frequency at the mid-point of the spectrum of the baseband waveform. Since this has components extending from $\omega_{\max}$ to $\omega_{\min}$, the frequency of this sinusoid is:

$$\omega_1 = \frac{\omega_{\max} + \omega_{\min}}{2} \tag{6.6}$$

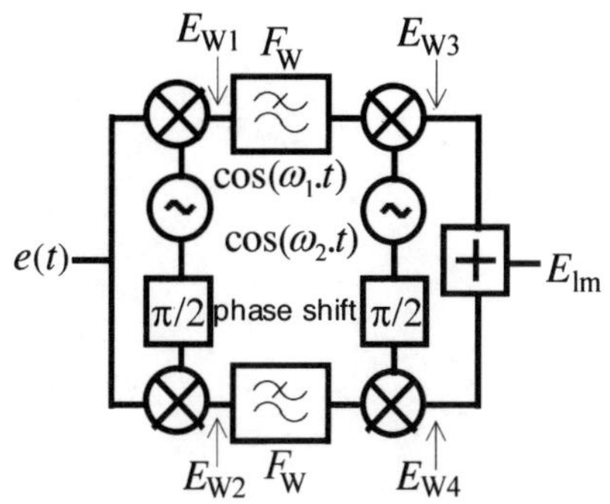

Fig. 6.4
The third (Weaver) method of generating LM.

Thus the result of this multiplication in channel (1) is:

$$E_{W1} = E \cdot \left[\sum \frac{c_r}{2} \{\cos([\omega_1 - \omega_r]t - \phi_r) + \cos([\omega_1 + \omega_r]t + \phi_r)\} \right] \tag{6.7}$$

The sine wave used for multiplying channel (2) is phase-shifted by $\pi/2$. The voltage in the second channel (with quadrature multiplication) is thus:

$$E_{W2} = E \cdot \left[\sum \frac{c_r}{2} \left\{ \cos\left([\omega_1 - \omega_r]t - \phi_r + \frac{\pi}{2} \right) + \cos\left([\omega_1 + \omega_r]t + \phi_r + \frac{\pi}{2} \right) \right\} \right] \tag{6.8}$$

All the components of the baseband signal have been transferred into the frequency range zero to $(\omega_{max} - \omega_{min})/2$. A low-pass filter is therefore used in each channel to clean up the signal. We define a 'brick wall' low-pass filter function F_W such that:

$$\left. \begin{aligned} F_W[f(\omega)] &= f(\omega) \quad \text{for} \quad \omega \leq \omega_1 \\ &= 0 \quad \text{for} \quad \omega > \omega_1 \end{aligned} \right\} \tag{6.9}$$

But since $[\omega_1 + \omega_r > \omega_1]_{\text{all } r}$ it follows that:

$$F_W[E_{W1}] = E \cdot \sum \frac{c_r}{2} \cos([\omega_1 - \omega_r]t - \phi_r) \tag{6.10}$$

Similarly:

$$F_W[E_{W2}] = E \cdot \sum \frac{c_r}{2} \cos\left([\omega_1 - \omega_r]t - \phi_r + \frac{\pi}{2} \right) \tag{6.11}$$

After low-pass filtering both E_{W1} and E_{W2} are multiplied by a second sinusoid at a frequency ω_2 (the first without carrier phase shift and the second with a phase shift of $\pi/2$). So let:

$$\begin{aligned} E_{W3} &= F_W[E_{W1}]\cdot\cos(\omega_2 t) \\ &= E\cdot\sum\frac{c_r}{2}\cos([\omega_1-\omega_r]t-\phi_r)\cdot\cos(\omega_2 t) \end{aligned}$$

and similarly:

$$\begin{aligned} E_{W4} &= F_W[E_{W2}]\cdot\cos(\omega_2 t) \\ &= E\cdot\sum\frac{c_r}{2}\cos\left([\omega_1-\omega_r]t-\phi_r+\frac{\pi}{2}\right)\cdot\cos\left(\omega_2 t+\frac{\pi}{2}\right) \end{aligned}$$

Converting the products of cosines into sums in the usual way:

$$\begin{aligned} E_{W3} &= F_W[E_{W1}]\cdot\cos(\omega_2 t) \\ &= E\cdot\sum\frac{c_r}{4}\{\cos([\omega_2+\omega_1-\omega_r]t-\phi_r) \\ &\qquad + \cos([\omega_2-\omega_1+\omega_r]t+\phi_r)\} \end{aligned}$$

also:

$$\begin{aligned} E_{W4} &= F_W[E_{W2}]\cdot\cos\left(\omega_2 t+\frac{\pi}{2}\right) \\ &= E\cdot\sum\frac{c_r}{4}\{\cos([\omega_2+\omega_1-\omega_r]t-\phi_r+\pi) \\ &\qquad + \cos([\omega_2-\omega_1+\omega_r]t+\phi_r)\} \end{aligned}$$

Suppose now that E_{W3} and E_{W4} are added, the components which have a phase shift of π radians (phase inversion) will cancel their counterparts which do not, leaving:

$$E_{W3}+E_{W4} = E\cdot\sum\frac{c_r}{2}\{\cos([\omega_2-\omega_1+\omega_r]t+\phi_r)\} \qquad (6.12)$$

Or, writing $\omega_c = \omega_2-\omega_1$:

$$E_{lm} = E_{W3}+E_{W4} = E\cdot\sum\frac{c_r}{2}\{\cos([\omega_c+\omega_r]t+\phi_r)\} \qquad (6.13)$$

Since all the components of the baseband waveform are present in E_{lm} but shifted up in frequency by an equal amount ω_c, evidently this is a sum-type LM waveform with notional carrier frequency ω_c. Had we taken the difference of E_3 and E_4 we would have obtained the difference LM waveform, with a notional carrier frequency $(\omega_1 + \omega_2)$. (The carrier frequency is notional because no carrier component is present in the LM waveform provided all $\omega_r > 0$.)

This rather complicated-seeming process for generating an LM signal is actually easier to implement than either of the other two. The low-pass filtering F_W takes place in the middle of the audio frequency range, and is therefore relatively easy to realize in analogue active filter form, whilst if it is implemented using digital processing the clock rate needed is substantially lower than for RF filters. What is more, if the cancellation of the out-of-phase components is imperfect the result cannot be any spurious outputs in the adjacent channel (as with the second, phasing method), but corresponds to a low-level signal superimposed on the wanted signal and identical with it except for a frequency-reversed spectrum. Subjectively this is perceived as a form of distortion, and is tolerable in communication-quality speech even when as high as -15 dB, a level easily achieved.

The only real disadvantage of the Weaver method is that because the filter cut-off is in the middle of the audio band, there is the possibility of phase and amplitude variation for signal components adjacent to this frequency. For analogue speech and even music, with care this need not be significant, but for other modulating signals it can be critical.

DSP algorithms based on all three methods of generating LM are widely used. The first and second (filter and phasing) methods are usually implemented at a relatively low notional carrier frequency (such as 15 kHz for communication-quality speech) with a subsequent frequency up-conversion. This reduces the clock-rate requirements both for digital filtering and Hilbert transforms, with consequent savings on DSP chip power consumption – particularly important for battery powered equipment. Design problems are most tractable if the carrier frequency is kept as low as possible, but not less than four times the highest modulating frequency.

6.4 Demodulating the LM signal

In principle the demodulation of LM signals is very simple. Taking the waveform of (for example) equation (6.13) and multiplying by a sinusoid near the nominal carrier frequency:

$$E_{\mathrm{lm}}\cdot\cos([\omega_{\mathrm{c}}+\Delta\omega]t+\phi_{\mathrm{c}})$$

$$=E\cdot\sum\frac{c_{\mathrm{r}}}{4}\{\cos([2\omega_{\mathrm{c}}+\Delta\omega+\omega_{\mathrm{r}}]t+\phi_{\mathrm{r}}+\phi_{\mathrm{c}})$$

$$+\cos([\omega_{\mathrm{r}}-\Delta\omega]t+\phi_{\mathrm{r}}-\phi_{\mathrm{c}})\} \quad (6.14)$$

A low-pass filtering function F_{d} is applied, which cuts off above the highest modulating frequency and we obtain:

$$e'(t)=F_{\mathrm{d}}\{E_{\mathrm{lm}}\cdot\cos([\omega_{\mathrm{c}}+\Delta\omega]t+\phi_{\mathrm{c}})\}$$

$$=E\cdot\sum\frac{c_{\mathrm{r}}}{4}\{\cos([\omega_{\mathrm{r}}-\Delta\omega]t+\phi_{\mathrm{r}}-\phi_{\mathrm{c}})\}$$

In the particular case when the locally generated sinusoid is exactly at the correct carrier frequency and in-phase with it:

$$\Delta\omega=\phi_{\mathrm{c}}=0$$

so:

$$e'(t)=E\cdot\sum\frac{c_{\mathrm{r}}}{4}\cos(\omega_{\mathrm{r}}t+\phi_{\mathrm{r}})=\tfrac{1}{4}\cdot e(t) \quad (6.15)$$

The original modulating function has been correctly recovered. This is the ideal case of **coherent demodulation**. To achieve it the locally generated sinusoid must be phase-locked to the notional carrier in some way. There are many ways in which this can be done, but the commonest involve transmitting an additional continuous sinusoid of constant amplitude with the LM transmission. Since the frequency of this component is known, it can be extracted by narrowband filters even if its amplitude is so low that its contribution to spectrum pollution or waste of transmitter power are both negligible. Typically **pilot tone** transmissions of this kind are at least 13 dB (20×) lower in amplitude than a full AM carrier, so that interactions between two such tones is down at least 26 dB compared with AM, making them almost insignificant.

6.5 Coherent LM transmissions

Transmission of a continuous pilot tone is one of the commonest ways of synchronizing LM transmissions. There are three variants depending on where in the signal spectrum the tone frequency is chosen to be: **tone below band (TBB)**, **tone above band (TAB)** and **tone in band (TIB)**. (The LM signal is here assumed to be in the sum form, but exactly comparable arguments can be developed for the less-common difference form.)

The only TBB system to have seen serious use places the tone at the nominal carrier frequency; it is often called a **pilot carrier** system and its close relationship to DSBDC is obvious. In the receiver a local RF oscillator is phase-locked to the pilot carrier, using a phase-lock loop, and can then be used for coherent demodulation. This seems simple enough, but has the serious disadvantage that the RF phase-lock loop may operate at an inconveniently high frequency for cheap and simple hardware realization, particularly as strong modulation sidebands are only two or three hundred hertz away from the pilot.

By contrast, in TAB and TIB systems the pilot tone is extracted at AF after the demodulator, which is much easier. How is it used to generate a coherent local carrier? In modern radio transmitters and receivers, the locally generated carrier will derive from a frequency synthesizer (Fig. 6.5). This uses a precise quartz crystal clock oscillator, and derives all necessary frequencies from it by digital means, using counters and phase-lock loops. Thus if the frequency standard provided by the clock oscillator is ω_S then it is easy to arrange that:

$$\omega_c = \frac{m}{n} \cdot \omega_S \quad \text{and} \quad \omega_t = \frac{1}{p} \cdot \omega_S \tag{6.16}$$

where m, n, p are integers and ω_t is the tone frequency.

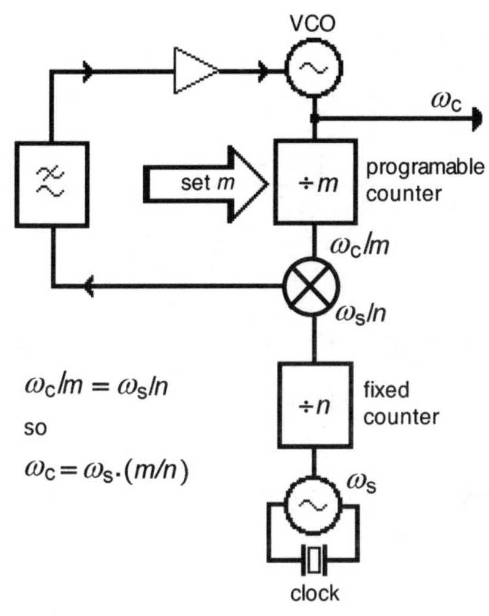

Fig. 6.5
A frequency synthesizer, using counters and phase locking to relate the generated frequency integrally to a clock frequency.

For example, in a TAB system if the clock frequency were nominally 5 MHz, the carrier frequency 100 MHz and the tone 5 kHz, then in both transmitter and receiver it would be enough to set $m = 20$, $n = 1$ and $p = 1000$. In the receiver, a phase-lock loop is used to 'pull' the frequency of the 5 MHz clock oscillator very slightly, so that the locally generated tone is phase-locked to the received tone (Fig. 6.6). It then follows that the local carrier must be similarly phase-locked, and can be used for coherent demodulation. In practice the clock oscillator will usually be a **voltage-controlled crystal oscillator** (**VCXO**) to permit this very fine frequency adjustment or 'pulling'. By this process of frequency locking, the demodulation is made coherent with the transmitted signal, so that the received baseband signal is virtually identical with the original.

TIB systems are phase-locked exactly like TAB systems. In the example of the previous paragraph, if p had been 3000 the tone frequency would have been 1.667 kHz, a typical value for a TIB system; phase coherence is established in the same way. However, TIB systems put the pilot tone in the middle of the channel, so if

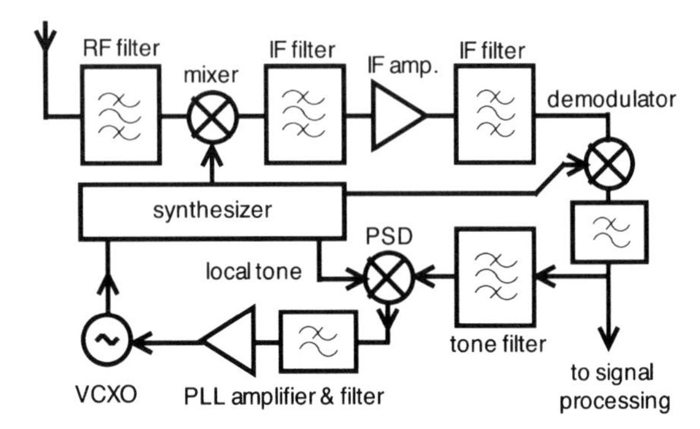

Fig. 6.6
Phase locking using the pilot tone makes possible coherent LM demodulation.

special precautions were not taken it would be heard as a loud whistle in analogue systems, and could cause errors in digital transmissions. Additional circuits have to be included in the TIB demodulator to overcome these problems. Yet in present-day practice TIB has quite driven out TAB, and also TBB. So what is the advantage of TIB over the apparently simpler alternatives?

6.6 TIB, TBB and TAB; the comparisons

The pilot tone in coherent LM systems is included for two distinct reasons. We have already analysed its use to establish frequency and phase coherence in the demodulation process, but it is also employed as an amplitude reference. Since it is transmitted at a constant amplitude, if the operation of the receiver automatic gain control (AGC) system is arranged to hold the demodulated pilot-tone amplitude constant it follows that the overall transmission channel gain is invariant, despite any fading on the radio link. This can be achieved using the familiar feedback AGC systems, although if very fast fading is expected it is possible that feedforward AGC will also be required. Both functions of the pilot tone are facilitated if it is placed at the centre of the transmitted signal (TIB) rather than at the edge (TBB, TAB).

We begin with phase and frequency locking. Initially the receiver locally generated frequencies will be offset from the desired value as a result of a small error in the local clock frequency, due to crystal secular drift and temperature sensitivity. In low-cost systems this drift may exceed ±3 parts per million and will rarely be less than ±1 ppm, so that, for example, at a carrier frequency of 450 MHz the drift would equate to ±1.35 kHz in cheaper equipment. This would be within the passband of a TIB receiver for both directions of drift, but with TBB and TAB systems would be outside the passband in one direction, so that in this case frequency lock would never be established because the pilot would not be received. This problem – obviously more severe at higher carrier frequencies – leads to a powerful argument in favour of TIB. TBB and TAB systems can be made to work, for example by sweeping their carrier frequency over a range and detecting when they come within the receiver pass-band (so-called **scan-lock systems**) but they loose the simplicity which is their major asset.

In fast-moving vehicles, such as aircraft, missiles and space probes, there is an additional frequency offset between ground-based equipment and that on the vehicle due to **doppler shift**. For example, this exceeds 2 ppm in the Concorde supersonic airliner at cruising speed. Here too TIB is at a marked advantage for similar reasons.

So far as amplitude equalization is concerned, the principal difference between the systems becomes obvious when amplitude fluctuations due to channel fading are fast. In this case it is essential that the fading of the pilot tone and the sidebands shall be as closely correlated as possible, otherwise the correction of amplitude fluctuations by AGC will be imperfect. Putting the pilot tone in the middle of the channel maximizes this correlation for two important reasons.

The first arises because the fading of a radio channel (arising from multipath effects) is not frequency independent. If a radio signal arrives at the receiver by two independent paths, the signals received by the two paths will add if the phase difference in path lengths is zero or an even multiple of π radians, but will subtract if it is an odd multiple. However, the phase difference is just:

$$\Delta\phi = 2\pi \cdot \frac{\Delta r}{\lambda} = \frac{\Delta r}{c} \cdot \omega \quad \text{where} \quad \Delta r = \text{range difference} \tag{6.17}$$

Since this phase difference is frequency dependent, it follows that there will exist frequencies ω_{max} at which the signals add and others ω_{min} at which they cancel given by:

$$\omega_{max} = \frac{2n\pi \cdot c}{\Delta r} \quad \text{and} \quad \omega_{min} = \frac{(2n+1)\pi \cdot c}{\Delta r} \tag{6.18}$$

The difference between adjacent maxima and minima frequencies is the **coherence bandwidth**, given (in Hertz) by:

$$\Delta f_{colt} = \frac{(\omega_{min} - \omega_{max})}{2\pi} = \frac{c}{2\Delta r} = \frac{1}{2\Delta t} \tag{6.19}$$

Here Δt is the difference in transit time on the two radio paths.

Thus, for multipath propagation in the VHF and UHF terrestrial environments, these variations in transit time range from a fraction of a microsecond at the higher frequencies to a few microseconds at the lower. This corresponds to coherence bandwidths from a few hundreds of kilohertz down to a few tens of kilohertz. If fading of a pilot tone is to be closely correlated with that of a sideband they need to be separated in frequency by an amount which is a very small fraction of the coherence bandwidth. The crucial case is when the received sidebands over two paths are very nearly in antiphase, and thus almost cancel. Because we are dealing with the small difference between large variables, very small deviations between the fade status of the sidebands and the pilot will produce marked results. For example, in a 20 dB deep sideband fade the two interfering signals will be opposed and within 1% of the same amplitude. Should the pilot, at a different frequency, derive from two components which are within 2% of each other the resultant will be twice the size of that for the sidebands, and adjusting the system gain to bring it to a standard value will still leave the sideband at half the amplitude it ought to have.

A simple calculation demonstrates that to hold the required amplitude even to 10% of its notional value requires that the frequency difference between the pilot and sideband should be under 3% of the coherence bandwidth – from a few kilohertz down to a few hundred hertz in the examples of VHF and UHF terrestrial radio already quoted. For TIB the separation is (at worst) half that for TAB, so TIB will always give better compensation for fading, adequate for the environment considered, although only just so at the VHF extreme.

The second advantage of TIB becomes apparent when the fading is considered in the time domain. To correct fading of the sidebands by varying gain so as to compensate the fading of the pilot tone requires crucially that there should be no difference in time delay between tone and sidebands, right through the system. However, all radio receivers incorporate a band-pass filter of some kind to select the required channel from among others. Usually (but not always) this is the IF band-pass filter, formerly an LC or crystal filter but now increasingly realized in digital form. In either case, the filter invariably has Chebyshev or similar response, because a very rapid rise in attenuation is required beyond the edges of the pass-band. However, this extreme change in insertion loss at the band edges corresponds to a greatly increased time delay compared with that at the centre of the passband (Fig. 6.7). But with TBB or TAB systems the region of maximum delay is precisely where the pilot tone is located, and as a consequence it is impossible for AGC systems to compensate the channel-derived amplitude variations. This problem, which is more intractable the more rapid the fading, is

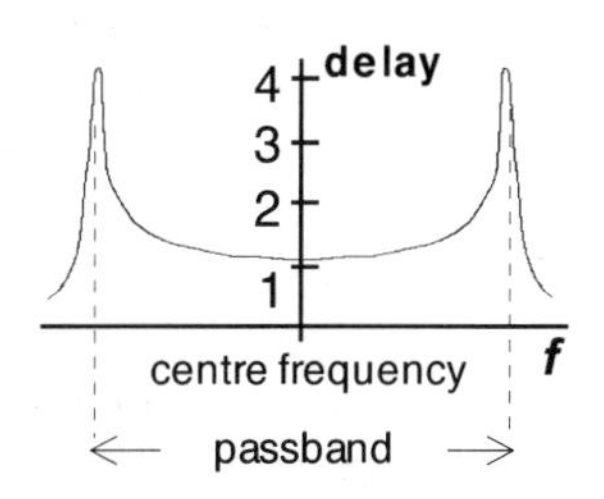

Fig. 6.7
Time delay in a Chebyshev band-pass filter (10 pole, 0.1 dB ripple in the passband). (Arbitrary vertical scale)

completely overcome by placing the pilot near the middle of the passband, as with TIB.

6.7 Implementing TIB LM systems

For LM transmission the advantages of TIB are so clear that TBB and TAB may now be considered as obsolescent. But how is the effect of the pilot tone in the received signal overcome? A continuous tone in the middle of the signal frequency band, even at low level, would be quite intolerable.

Early systems simply band-stopped a few hundred hertz near the middle of the AF spectrum, placing the pilot in the spectrum 'hole' so produced. It is a surprising fact that in speech a spectrum 'hole' of this kind, even if two or three hundred hertz wide, has a hardly measurable effect on speech intelligibility or naturalness, provided only that it falls between 1 and 2 kHz. Even for music, band-stopping in this way has far less audible effect than might be expected, largely because musical tones are far from being pure sinusoids but are mostly rich in harmonics. If a single harmonic component, or even the fundamental, is removed by filtering, the presence of other harmonics results in the brain inferring the missing component. In the receiver, once the pilot has been extracted for control and equalization purposes, the AF signal is simply band-stopped once more so that the tone is inaudible to the listener.

Simple TIB systems of this kind do not find wide acceptance, because of the constraints that they place on the kind of signals that could be transmitted. For example, it is commonplace to wish to transmit data over voice channels using a modem, but in the presence of an AF band-stop this is very unsuccessful, with unacceptably high error rates. However, all the difficulties of implementing TIB were overcome with the introduction of **transparent tone-in-band (TTIB)**.

In the TTIB system the AF spectrum is split in two, and the upper part is translated upward, typically by 300 Hz (Fig. 6.8). This leaves

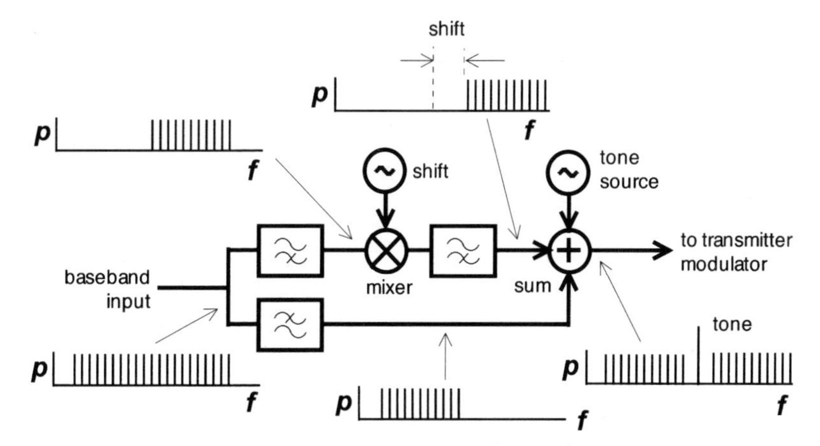

Fig. 6.8
Signal processing for a TTIB LM transmitter. Usually instantiated in DSP form.

a modified AF spectrum with a gap near its centre into which the pilot tone can be introduced. The signal is transmitted in this form. At the receiver the pilot is extracted in the usual way and used to equalize the channel. The upper part of the frequency spectrum is translated downward in frequency and reassembled phase coherently with the lower part, thus restoring the original AF spectrum without impairment. A channel of this kind is transparent to any signal type that falls within the available bandwidth, hence the name.

The complexity of the signal processing required discouraged the use of TTIB in the epoch of analogue circuits, but now that digital signal processing is available at low cost this objection no longer applies. TTIB LM systems are now finding growing acceptance, and as spectrum congestion gets worse this trend may be expected to continue.

6.8 Near-LM systems

Although LM, requiring the radiation of a single set of sidebands at the transmitter and their coherent demodulation at the receiver, approaches closely to the theoretical ideal for radio transmission, until the arrival of digital processing the complexity of the equipment needed to realize this approach led to unacceptable cost levels. In consequence less than perfect, but simpler, solutions to the problem of more effective spectrum use have been preferred in the past.

Analogue television broadcasting began using AM, but because the **video frequency (VF)** spectrum extends from zero frequency up to several megahertz, the fact that the RF bandwidth of an AM signal was twice this upper limit proved an early problem. Not only was it difficult to make frequency assignments of adequate width, but also the receiver design was made much more difficult, because of higher noise levels and poorer amplifier gain. This led to the introduction of **vestigial sideband (VSB)** transmission, which goes a good part of the way towards the LM ideal, yet is relatively simple.

Generating a VSB signal is quite like generating an LM signal using the filter method; in both cases the starting point can be a full AM signal, and a filter is used to attenuate the carrier and one set of sidebands. However, in the case of VSB, the filter is designed to have a relatively gradual roll-off, so that instead of the carrier and one sideband set being suppressed entirely, the carrier is in fact reduced to half and some nearer components of the unwanted set of sidebands are allowed to remain in progressively attenuated form (Fig. 6.9).

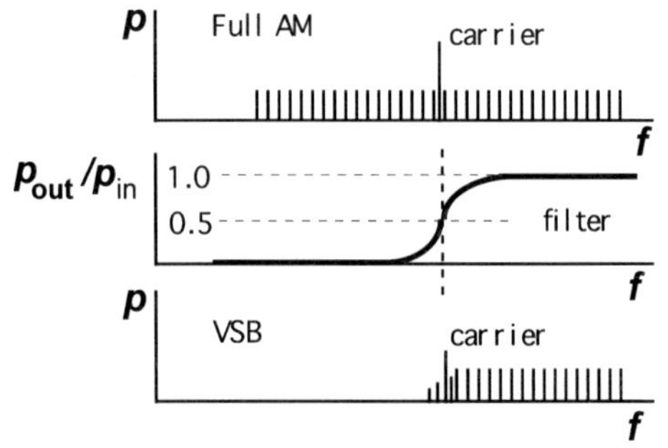

Fig. 6.9
Generating VSB from full AM requires only an undemanding filter.

Two results follow. The first is that a filter having this kind of characteristic can be simpler and cheaper than one which rolls off very sharply in the stopband. This eases transmitter construction. A few early VSB television transmitters actually generated an AM signal at full power level and filtered at high power (dumping the unwanted energy of the carrier and sidebands into a water-cooled load), but this was soon superseded by the more economical procedure of generating the VSB signal at relatively low level, followed by a linear amplifier to take it up to the desired transmission power.

Secondly, because a significant carrier component remains, the envelope of the wave is still a (somewhat distorted) representation of the original baseband modulating signal. This makes it possible to use a conventional rectifier demodulator (as with full AM), simplifying the receiver. True, there is some non-linear amplitude distortion, but this is tolerable in television systems, if not too large, since it corresponds only to a compression of the grey-scale in parts of the picture, to which the eye is not very sensitive. In 'positive modulation' systems, where maximum signal corresponds to white, there is some reduction in highlight detail, while in the alternative arrangement (maximum modulation equals black) shadow detail is lost; in either case the effect is not large and can be offset by modest predistortion.

In the late 1930s when VSB came into vogue for television the simplicity of both receiver and transmitter hardware compared with any form of LM then conceivable made the system seem an inevitable choice. Once VSB became established the investment in it soon grew too large for any change to be contemplated until the coming of digital television, some sixty years later.

Digital signals are also not very sensitive to amplitude distortion, so VSB also saw some use for data transmission, although it has now been overtaken by more modern approaches. For speech and music transmission VSB has never proved satisfactory because the non-linear amplitude distortion it introduces gives rise to unpleasant audible effects. The spurious harmonics produced prove unacceptable to listeners, even on relatively low-quality communication circuits.

As with VSB in television, the key to simple solutions for audio transmission has been to take advantage of the insensitivity of the human observer to certain kinds of signal impairment. For speech, although the hearer is intolerant of more than a very few per cent of amplitude distortion, there is an alternative psycho-acoustic phenomenon which can be exploited. In most forms of human speech the phase relationship of the various Fourier components of the AF waveform can be greatly in error without any measurable reduction in the intelligibility of the speech. Indeed it is sometimes said that the human ear has no sensitivity to phase. This remark must be treated with caution, since obviously the waveform of a transient sound, such as a sharp click, depends for its shape on the phase relationship of its Fourier components; all hearing people can tell the difference between a sharp click and a dull thud. Fortunately only a very few human languages are built around sharp clicking sounds, so for most speech the ear can indeed be treated as phase insensitive. Thus coherent demodulation of a sideband set is not essential.

Even a slight frequency error in the demodulated components of the AF wave is surprisingly well tolerated. This will happen if the locally generated sinusoid used to multiply the received sidebands has a small frequency error. In this case the baseband modulating waveform:

$$e(t) = \sum_{\text{all } r} c_{\text{r}} \cos(\omega_{\text{r}} t + \phi_{\text{r}})$$

is replaced by a demodulated waveform:

$$e_{\text{d}}(t) = \sum_{\text{all } r} d_{\text{r}} \cos\{(\omega_{\text{r}} + \delta\omega)t + \psi_{\text{r}}\} \qquad (6.20)$$

Subjective testing demonstrates that there is little loss of intelligibility of speech for a frequency error (in hertz) between -50 and $+30$ hertz.

This simple psycho-acoustic fact made practicable an early and widely used pseudo-LM system known as **single sideband (SSB)**. The transmitted signal is similar to LM without a pilot tone, but at the receiver it is demodulated non-coherently using a local oscillator close to, but not necessarily exactly at, the carrier frequency. Often means are provided to adjust the frequency manually over a narrow range for best audible effect; such a control is known as a **clarifier**. SSB has been (and still is) widely used in the HF band (3–30 MHz) where the frequency accuracy required of local oscillators (30–50 Hz) is possible with cheap crystal oscillators. Use of a clarifier may be avoided altogether, or at least it need only be adjusted at infrequent intervals.

SSB is a 'cheap and cheerful' speech-only system, and gives good results in that role. It is almost useless for music, where the frequency tolerance has to be at least ten times better to give barely passable quality. Data transmission by SSB has seen some use, but far better performance results with a coherent LM system, with orders of magnitude less error for the same S/N ratio.

Questions

1. A baseband signal has a spectrum extending down to zero frequency. How could an LM RF signal be generated from it? (only by the third method) From what principal impairment would the LM signal suffer? (superimposed low-level reversed spectrum)

2. Why do synthesizers take a time to settle after a frequency change which is roughly proportional to the inverse of the channel spacing?

3. An LM system conveys data signals having a baseband spectrum extending from 300 Hz to 3 kHz. What is the optimum pilot tone frequency and in the case of multipath propagation, what is the maximum path length difference which will ensure that no signal component is reduced by more than 10% in amplitude? (1.65 kHz, 18.8 km)

CHAPTER 7

SPECTRUM CONSERVING DIGITAL TRANSMISSION

Although, worldwide, analogue radio transmission is still used for most services, digital radio is taking over rapidly. Pioneered in the 1930s and having its roots in the telegraph transmission with which radio began, the advantages of digital signals are now recognized to be so large that we may reasonably expect that in due course analogue radio transmissions will disappear entirely, although doubtless that must take some considerable time. Transmitting digital signals in a way that conserves the electromagnetic spectrum to a maximum degree is therefore rightly seen as a problem of the utmost importance. The key to its solution is the radiation of a very strictly controlled and optimized spectrum from the transmitter.

Two factors are important in achieving an optimum transmitter radiation profile. The first is the choice of the system of RF modulation, by which the baseband signal is translated into the correct frequency band and assigned channel. The second is optimizing the form of the baseband signal derived from the digital input; this is done by the **modem** (a potentially confusing term derived from '**mod**ulator–**dem**odulator' but having nothing necessarily to do with RF modulation).

The RF modulation problem can be dealt with briefly. For digital baseband signals, just as with analogue, from the point of view of spectrum conservation the optimum system of RF modulation is coherent LM. CDM, to be discussed later, is also highly effective but may be considered as a variant of LM in which the carrier frequency is generalized to **sequency**, its hyperdimensional counterpart. The only important difference between analogue and digital LM is that it may be possible to dispense with the pilot tone in digital systems if the baseband signal itself has a component which can be used instead. All other systems of modulation are inferior to LM, but both AM and FM have been used in the past because of the availability of equipment and from reasons of commercial and technological inertia.

7.1 The modem problem

The design of the modem, which conditions the digital signal to a form suitable for transmission, is a complex matter. The source of digital signals for transmission is commonly either a computer or a **codec**, which converts analogue speech, music or television signals into digital form. They appear at the radio equipment in the form of clocked pulse trains, essentially rectangular waveforms. Mostly they are **unipolar**, consisting of transitions between just above 0 volts and near 3.3 or 5 volts (or some other fixed value). They may sometimes be **bipolar**, switching between positive and negative voltages; the difference is trivial, amounting to no more than the addition of a constant.

As is well known, a square wave of period T has an infinite spectrum of odd harmonics, and is of the form:

$$y(t) = \sum_{0}^{\infty} \frac{1}{(2r+1)} \cos\{(2r+1)\omega t\} \quad \text{where} \quad \omega = \frac{2\pi}{T} \tag{7.1}$$

Using such a waveform directly to modulate the transmitter would result in an infinitely wide spectrum, although with sidebands declining in amplitude with remoteness (in frequency) from the carrier. This is obviously unacceptable since sidebands with powers as low as −60 dB can cause significant adjacent channel interference in many radio systems, the falling amplitude of the higher sidebands is no solution to the problem: interference potential would remain out to the 999th harmonic!

Another difficulty of the simple rectangular wave is that it also has some very low frequency components. A sequence of the same digit – **0**s or **1**s – will result in a steady voltage over a prolonged period, and such sequences cannot be ruled out. However, most RF modulation schemes (other than VSB) require a lower as well as an upper limit on the baseband spectrum, for example in the case of LM so as to allow one sideband set to be filtered out using realizable filter characteristics. Generally modulators accept signals within a well-defined baseband frequency range, 300 Hz to 3 kHz for communication quality speech and wider for other purposes. Modems are designed to process the digital input to produce an output having a spectrum entirely contained within this bandwidth. How is this done?

Looking at a waveform in the frequency domain, we note that the higher frequency spectrum components are not essential to communicating the digital information since a train of pulses (supposing for the moment that they are bipolar) is fully defined by the zero crossings. In between zero-crossings the wave can take any shape without modifying the significance of the digital signal; for present purposes it is obvious that the transmitter spectrum will spread least if the waveform approximates as closely as possible to a sinusoid.

The obvious way to remove the higher harmonics might seem to be by low-pass filtering, but this approach has serious problems. To remove the third and higher harmonics a filter would be required to cut off above the fundamental but introduce a large stopband attenuation at three times that frequency. This is a rapid roll-off, and filters having this characteristic in the frequency domain have a very poor transient response in the time domain. Orders of filters which have a sharp 'corner' on their frequency domain response curve, such as the Chebyshev, achieve this by introducing some kind of resonance near the cut-off frequency which sustains the response fairly uniformly until it drops sharply. When subjected to a sharp input transient waveform the resonant components are shock excited into a damped oscillation – the filter 'rings' (Fig. 7.1, based on Zverev (1967), still the definitive text on the subject). This continued output confuses subsequent signals and it proves impossible to decode without error a digital signal which has passed through such a filter. By contrast those orders of filter, such as the

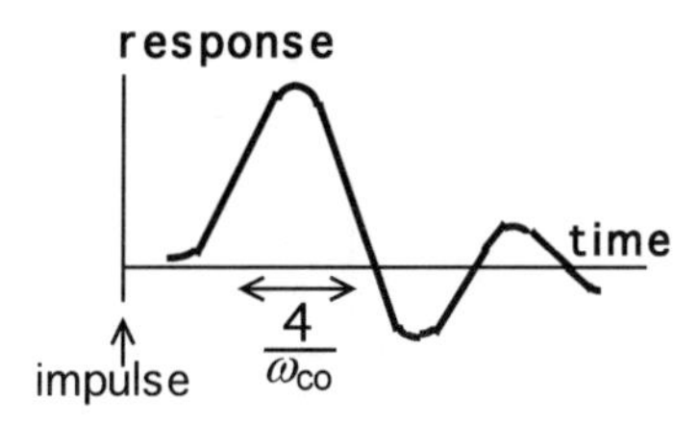

Fig. 7.1
Impulse response of a Chebyshev low-pass filter (5 pole, 0.5 dB ripple).

Gaussian, which have a well-behaved transient response also have a very gentle roll-off in the frequency domain, and are therefore inadequate to suppress the lower harmonics. So low-pass filtering can make a contribution to shaping the digital waveform by suppressing the higher order harmonics, but something else is also required.

The solution is to replace the typical rectangular digital waveform with another wave shape, having a more acceptable spectrum. Such shapes can be stored digitally, requiring very little memory (RAM, or more usually ROM). For example, consider a digit waveform defined by:

$$\begin{aligned} y_{\mathrm{d}} &= 0 \quad t < 0, t > T \\ &= 1 \quad 0 < t < T \quad \text{for a } \mathbf{1} \text{ digit} \\ &= 0 \quad 0 < t < T \quad \text{for a } \mathbf{0} \text{ digit} \end{aligned} \tag{7.2}$$

This is a rectangular pulse having a clock rate $1/T$ and with the usual disadvantages of high harmonic content. Suppose now that it is replaced by:

$$\begin{aligned} y_{\mathrm{rc}} &= 0 \quad t < 0, t > T \\ &= \frac{1}{2}\left\{1 - \cos\left(\frac{2\pi t}{T}\right)\right\} \quad 0 < t < T \quad \text{for a } \mathbf{1} \text{ digit} \\ &= 0 \quad 0 < t < T \quad \text{for a } \mathbf{0} \text{ digit} \end{aligned} \tag{7.3}$$

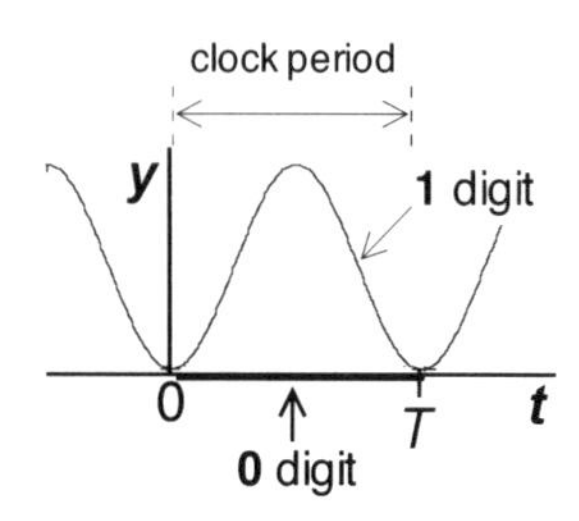

Fig. 7.2
A simple 'raised cosine' waveform. A sequence of **1**s is shown.

The waveform now has the shape of a cosine function, but with a constant added so that it just touches zero at its lowest point; it is known as a **raised cosine** (Fig. 7.2). Such a waveform is easily generated. A sequence of instantaneous amplitude values at regular intervals is stored for the raised cosine; these are subsequently read and equispaced in the time interval between successive zeros of the rectangular wave, then digital-to-analogue converted. A low-pass filter with good transient response is next used to remove higher frequency noise components due to the finite number of waveform samples. The more samples used to represent the waveform the less demanding the filter specification. Storage costs are negligible, so the limit on number of samples is determined by the rate at which they must be accessed during the period of one digit, T.

The spectrum of raised cosine waves is very simple: nothing at all on a sequence of **0** digits, a sinusoid at a single frequency $f = 1/T$ on a string of **1**s, and a more complex low-frequency spectrum on a mixed sequence.

In this simple form a string of **1**s gives rise to a baseband sinusoid of frequency equal to the data bit rate. However, using raised cosines in this simple form, although seen in designs of early date, is not optimal. It is far better only to ramp up (following the first half of the raised cosine form) at the beginning of a sequence of **1**s and to ramp down only at the end, so that the sequence is represented in the smoothed waveform as a 'plateau' with smoothly rising front

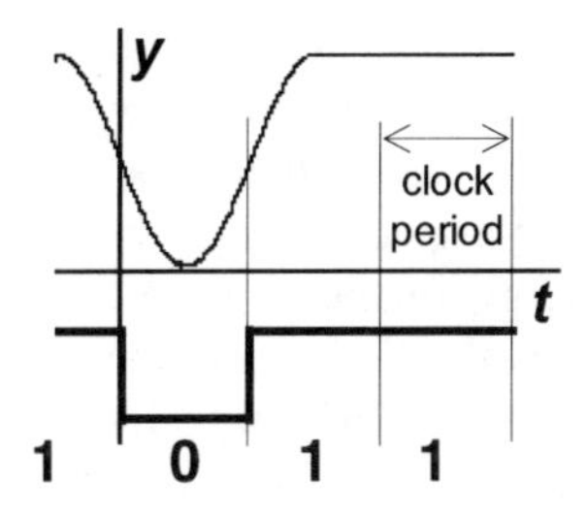

Fig. 7.3
A more sophisticated 'raised cosine' waveform; the clock period is halved for the same rate of change, doubling the possible data rate.

and back profiles – a **non-return-to-zero (NRZ)** waveform. This can be done by a look ahead at the next digit, which in practice means slightly delaying the digit stream by holding it briefly in a cache memory while this check is made.

A further refinement is to begin the ramp up for a **0–1** transition halfway through the **0**, completing it halfway through the **1**, and thus taking twice the time over it, compared with the simpler case where nothing happens until the beginning of the **1** digit (Fig. 7.3). So for a **0–1** transition at time 0:

$$y_{\mathrm{rc}} = \frac{1}{2}\left\{1 - \cos\left(\frac{\pi t}{T} + \frac{\pi}{2}\right)\right\} \quad \text{for} \quad -\frac{T}{2} < t < \frac{T}{2} \tag{7.4}$$

while for a **1–0** transition at time 0:

$$y_{\mathrm{rc}} = \frac{1}{2}\left\{1 - \cos\left(\frac{\pi t}{T} - \frac{\pi}{2}\right)\right\} \quad \text{for} \quad -\frac{T}{2} < t < \frac{T}{2} \tag{7.5}$$

This slowing down scales all the Fourier components down in frequency by a factor of two. Looked at another way, to keep the same Fourier components the clock period would be halved. Similarly the **1–0** transition begins halfway through the **1** (as before) but is completed halfway through the following **0**. Provided that we sense the state of the digit at its middle, the discrimination between **0** and **1** is no worse than before.

Using this more sophisticated waveform conditioning strategy, the highest frequency is now generated by a **0–1–0–1** . . . sequence and is equal to half the bit rate. Other, less regular, digit sequences can be shown to generate a spectrum largely bounded by this limit, though with a little energy further out. A rate of two bits per hertz of bandwidth is thus the theoretical maximum for a simple two-state baseband waveform of this kind. Once again, sampling noise in the output is cleaned away by using a low-pass filter of good transient response.

This argument has been developed using a cosine form as the 'smooth' shape used to reduce the signal bandwidth, compared with rectangular shapes, and raised cosine waves have indeed been widely used. Since the actual form will be stored in memory as a sequence of samples, there is no cost disadvantage in adopting an alternative shape. Smooth shapes other than a raised cosine could therefore also be used – a Gaussian (bell-shaped) waveform has been widely adopted because it shows a modest but worthwhile advantage in having fewer components beyond half the bit rate.

Either way, the digital waveform has now been conditioned so as to put a moderate upper limit on its spectrum; however, this still extends down to zero frequency, and is therefore not suitable for direct modulation on most radio transmitters (or transmission by telephone lines or cables, for that matter) all of which require a non-zero lower limit to the spectrum. It is therefore necessary to translate the spectrum of the conditioned waveform upward, a process of modulation. In principle it might be thought possible to use LM for this purpose, but the technology has not developed in that direction. Instead, equally effective modulating strategies have been developed, peculiar to data modems, which prove easier to implement. The digit stream is modulated on to a low frequency carrier, and the resulting signal, having a spectrum with upper and lower bounds, can then be modulated on to the RF carrier, optimally using LM. For example, in the land mobile service the data modem generates an analogue spectrum extending from 300 Hz to 3 kHz, which is then linearly modulated onto an RF carrier at several hundred megahertz. Because the process of LM is entirely transparent, this is equivalent to modulating the data directly on to the RF carrier, but much easier to do.

7.2 Phase-shift keying

How is the smoothed data waveform modulated on to the low frequency carrier in the data modem? Historically, the same methods were used as for analogue signals – AM and FM – although from an early stage waste of energy in the form of carrier components was avoided, so that the commonest AM-related form was DSBSC, known in the data world (for reasons which will become apparent) as **binary phase-shift keying (BPSK)**.

If the digital signal has a bit interval T the highest modulating frequency will arise from a **0–1–0–1** ... sequence and is a sinusoid of frequency $1/2T$ Hz. Using BPSK (DSBSC) modulation onto the carrier, the transmitted signal is:

$$E_{\mathrm{bpsk}} = E\cos\left(\frac{\pi}{T}t\right)\cdot\cos(\omega_{\mathrm{a}}t) \tag{7.6}$$

where ω_{a} is the low frequency carrier, or

$$E_{\mathrm{bpsk}} = E\cdot\left\{\cos\left(\left[\omega_{\mathrm{a}} - \frac{\pi}{T}\right]t\right) + \cos\left(\left[\omega_{\mathrm{a}} + \frac{\pi}{T}\right]t\right)\right\} \tag{7.7}$$

which may be compared with equation (5.12) above. The spectrum consists of two lines, one displaced above and one below the nominal carrier frequency by an amount (in hertz) equal to half the maximum bit rate. There is no carrier component. In the time domain the waveform envelope looks like the modulus of a sine function, but at each zero the phase changes instantaneously by π radians (180°).

One way of looking at the waveform is to see it as two phases, π radians apart, used to signal the two digit states – hence the name phase-shift keying' – but with the modulated envelope shaped so as to ensure that when the phase switching occurs the amplitude is zero. This gives a useful clue to the design of digital modulating waveforms which minimize spectrum pollution. Unwanted, polluting sidebands are capable of being generated when either the amplitude or phase of the modulated wave changes. Careful control

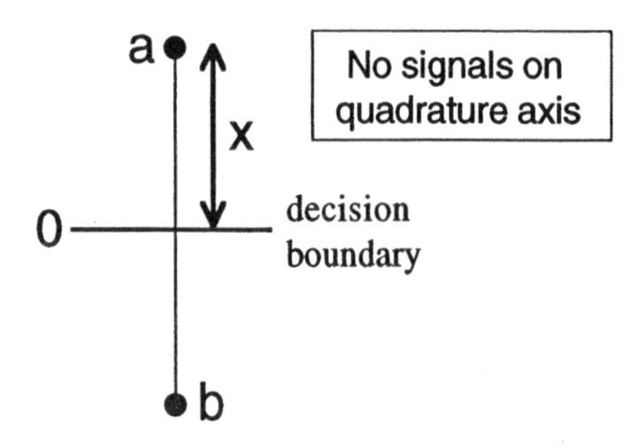

Fig. 7.4
BPSK represented in signal space.

of amplitude changes (in this case to make them conform to a sinusoidal envelope shape) minimizes the spectrum width due to this cause, while although it is true that the phase changes very rapidly, we arrange that it does so at a time when the amplitude is zero, hence avoiding non-zero sidebands due to phase modulation.

A useful alternative way to represent the BPSK signal is by means of a plot in polar co-ordinates, where the angle is the phase of the signal (relative to the nominal carrier) and the radius is the amplitude (Fig. 7.4). The result, in this case, is a vertical line extending above zero by the maximum amplitude, and below zero by the same amount but in antiphase. This is called a **signal space** plot. In a **1–0** transition the **trajectory** of the signal from its maximum value is down through zero (where an instantaneous π radians phase switch occurs) and out to maximum again. The points at the top and bottom of the trajectory (designated **a** and **b**) correspond to the extremes of the signal; it is here that S/N ratio is best, and hence the state of the digit can be checked with least risk of error. The horizontal axis is the **decision boundary**; above it the receiver modem will interpret the received signal as a **1** and below it as a **0**.

The BPSK waveform still has two sidebands. Why not use LM and halve the bandwidth needed? The answer is that a more convenient alternative strategy is available for digital signals which cannot easily be used at all in the analogue domain. The important point is that a stream of digits is easily split into two streams, each at half the data rate, and this gives crucial flexibility in system design. There are many strategies by which this splitting can be done, but the most common is for digital words (digit groups) to be routed alternately to the two channels and then re-timed at half the clock rate. Alternatively just the same may also be done with longer data blocks, or for some special situations it may even be that the more significant digits will be routed to one channel and the less significant digits to the other. Any of these strategies (and others) will generate two data streams, each with half the rate of the original.

Now consider these two digit streams $e_1(t)$ and $e_2(t)$, both conditioned to have minimal spectrum occupancy as described above. If each is modulated on a low frequency carrier as before, but with the carrier for one shifted $\pi/2$ relative to the other, the two signals become **orthogonal**. Although both are present in the receiver, they can easily be demodulated separately. Thus if the received signal:

$$E_r = e_1 \cos(\omega_a t) + e_2 \cos\left(\omega_a t - \frac{\pi}{2}\right) = e_1 \cos(\omega_a t) + e_2 \sin(\omega_a t) \tag{7.8}$$

then:

$$\begin{aligned} E_r \cdot \cos(\omega_a t) &= e_1 \cos^2(\omega_a t) + e_2 \sin(\omega_a t) \cdot \cos(\omega_a t) \\ &= \frac{e_1}{2} + \frac{e_1}{2} \cos(2\omega_a t) + \frac{e_2}{2} \sin(2\omega_a t) \end{aligned}$$

Applying the usual low-pass filter to extract the baseband component:

$$F(E_r \cdot \cos \omega_a t) = \tfrac{1}{2}\, e_1 \tag{7.9}$$

Similarly:

$$F\left(E_r \cdot \cos\left[\omega_a t - \frac{\pi}{2}\right]\right) = \frac{1}{2}\, e_2 \tag{7.10}$$

So the two digit streams can easily be extracted at the receiver, each in a way wholly independent of the other, by using quadrature demodulating sinusoids. Thus a waveform such as that in equation (7.8) carries both digit streams simultaneously without their interfering with each other. This type of data modulation is called **quadrature phase-shift keying (QPSK)**. It is one of the most popular and in widespread use. Allowing for the fact that two data streams are being sent together, two bits per hertz of bandwidth are transmitted, achieving the same spectrum utilization as would have been possible with LM and an undivided data stream; however, QPSK is easier to implement.

In signal space, the plot for QPSK identifies four signal maxima at $\pi/2$ radians (90°) spacing in angle, designated **a**, **b**, **c** and **d**, which might correspond to transmission of **11**, **10**, **00** and **01** respectively (Fig. 7.5). The decision boundaries are now two lines at right angles, rotated $\pi/4$ relative to the horizontal axis. Note that the distance from the signal points to the boundaries has been reduced by a factor of $1/\sqrt{2}(-3\,\text{dB})$; this is important and will be taken up later when the effects of noise are considered.

What are the optimal trajectories in the QPSK case? To go between **a** and **c** it might be thought necessary to pass through zero, as in BPSK, and similarly for transitions between **b** and **d**. However,

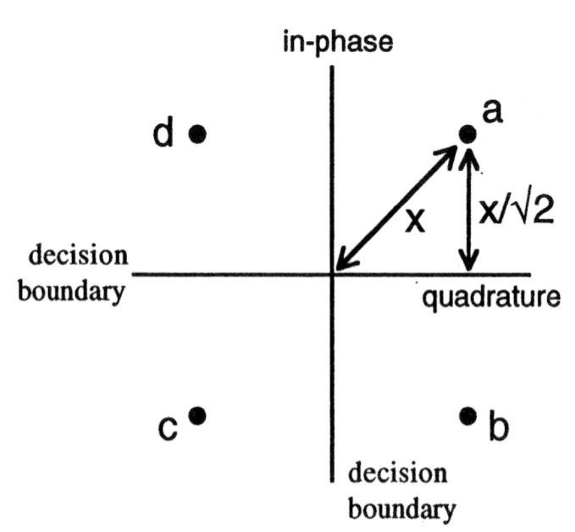

Fig. 7.5
QPSK represented in signal space. The same average power is split between in-phase and quadrature components.

there will always be a quadrature digit intervening between opposed pairs like this, so that provided that an **a** to **c** transition always takes the form **a** to **b** to **c** (or **a** to **d** to **c**) the amplitude of the modulated signal never drops to zero. The trajectories are the lines joining the points, and the amplitude falls by only 3 dB ($1/\sqrt{2}$) mid-way between them. This is a considerable advantage, because in practice it is quite difficult to build radio transmitters for BPSK in which the output can fall sufficiently close to a real zero during phase switching (and failure to do so will cause serious spectrum pollution), but for QPSK this problem need not arise.

There is one further complication. Do we constrain the changes always to correspond to movement in one rotational direction, for example clockwise, or do we allow rotation in either direction? The first may be easier to implement in hardware, but the second gives much better spectrum conservation. Consider an **a** to **d** transition; if only clockwise transitions are permitted this must involve a $3\pi/2$ (270°) phase change, whereas if anticlockwise changes are permitted it will be only $\pi/2$. In both cases the phase change must take place in the same time, so the rate of change of phase (which equates to instantaneous frequency) will be three times greater in the former case. This will correspond to a proportionately wider spectrum for the modulated signal. By allowing phase change in either direction we minimize the spectrum occupancy.

7.3 Noise and errors

Although QPSK can conveniently transmit data in as little bandwidth as LM, for this gain in speed there is a price to be paid. The distance in signal space between different digital codes has been reduced, increasing the possibility of error, should noise or interference be received along with the wanted signal. Before we can fully appreciate the merits and demerits of different possible systems, therefore, it is necessary to consider error rates in transmission. The transmission errors arise because each unique extreme signal state in signal space can no longer be represented by a point (as hitherto, two in the case of BPSK, four with QPSK). With the addition of noise, when averaged over time they are smeared out in all directions, so that they become patches of finite area, with ill-defined boundaries, corresponding to the probabilities of particular instantaneous locations. When these patches cross decision boundaries errors occur.

The theoretical analysis of error variation with interference given in the usual texts involves consideration of the error rates when the signal is mixed with noise, usually white or Gaussian noise. In fact in the present-day congested radio spectrum it is interference from other transmissions which causes the most trouble by far, but the mathematical analysis of this case is quite intractable because there are so many different possibilities to be considered. Looking at variation of error rate with noise is still thought useful; it gives us a helpful guide to the robustness of a system against interference, but little more. Derivation of the expressions for variation of error rate with S/N ratio for various data modulation systems is long and not particularly instructive, so it will not be reviewed here; we will simply consider the results obtained.

Because white noise is, by definition, uniform in power density over all frequencies, the actual noise power at the point where decisions about digits are made is itself proportionate to the bandwidth. The noise power is just the noise density multiplied by the bandwidth:

$$P_{\mathrm{n}} = \Delta\omega \cdot p_{\mathrm{n}}$$

where P_{n} is the total noise power and p_{n} the noise power density.

To simplify matters, therefore, we begin with the constant bandwidth case. For BPSK the variation of error rate is from 10^{-3} at +6.8 dB S/N power ratio, through 10^{-4} at 8.4 dB to 10^{-8} at 12 dB. The first thing to notice is that this is a very rapid variation in error rate with S/N ratio (Fig. 7.6). Something similar also turns out to be true of all the widely used modes of data modulation. Secondly, with many binary systems there would be very little gained from going beyond 10 or 12 dB S/N ratio, if this would only bring the transmission error rate far below other sources of error in the system, and might thus give no practical benefit. However, analogue speech circuits are commonly engineered to achieve S/N ratios of 20 or 25 dB; to use them for transmitting BPSK would be

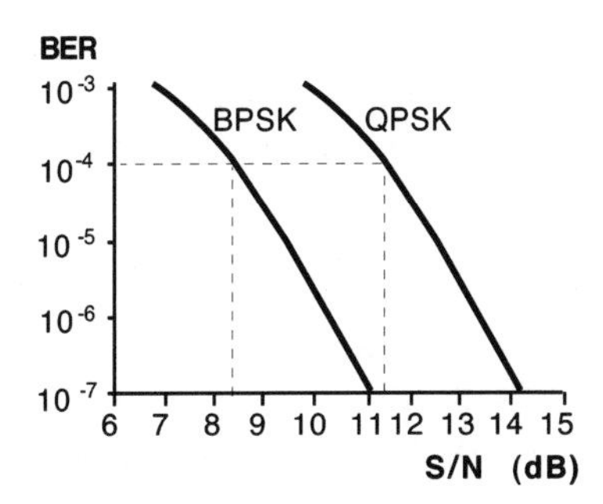

Fig. 7.6
Variation of raw bit error rate with S/N ratio.

wasteful of S/N ratio, and hence an inefficient use of the radio spectrum. For this reason a binary transmission system, like BPSK, would be used only where an unusually poor S/N ratio at the receiver was foreseen (and even here there are alternative strategies which may prove better, as we shall see). Examples are military systems, required to operate through heavy jamming, and extreme-range deep space probes, where for realistic transmitter powers the signals are bound to be weak due to length of the path over which they are transmitted.

To make the system more nearly optimal we have two options: in data-only systems we could directly reduce the potential for spectrum pollution by reducing the transmitter power, or we could use a more sophisticated data modulation system which exploits the better S/N ratio to achieve a higher transmission rate. The first has advantages where transmitter power is anyway very limited, such as in portable battery-powered equipment, but otherwise the higher-rate tactic usually gives a better outcome.

Consider the case of QPSK. Since, on a voltage plot in signal space the distance between the decision boundary and the signal points has been reduced by $\sqrt{2}(-3\,\text{dB})$, it would be expected that the noise must be scaled down by a similar factor to achieve similar error rates. The curve of error against S/N ratio for QPSK thus exactly follows that for BPSK, but shifted by 3 dB, so that for an error rate of 10^{-4} an S/N ratio of 11.4 dB is required (as against 8.4 dB in the binary case). However, so far we have been considering a fixed bandwidth, and with QPSK for a given bandwidth the data rate is, as we have seen, doubled compared with BPSK. We therefore have the option of staying with the original data rate and halving the bandwidth; for white noise this halves the noise power also. Thus the 11.4 dB S/N ratio required would be obtained for the *same signal power* as was needed to give the 8.4 dB S/N ratio in the binary case with wider bandwidth. What is more, we have saved half the spectrum that would have been used for BPSK, and (as already noted) QPSK transmitters with a clean spectrum are easier to build than BPSK.

These are good reasons to prefer QPSK; however, they do not always apply. The argument is built on the assumption that the interference encountered is white noise, or approximates to it reasonably well. This might not be valid, for example in the case of military jamming, which sometimes comes from a **follower jammer**, which is a narrowband frequency-agile RF source tuned by a control mechanism to sit on the centre of the channel occupied by the wanted transmission. Equally, some unintentional narrowband co-channel interference, often encountered in civil use, has similar characteristics. In situations of this kind the wider bandwidth does not result in more interfering power being received, so binary modulation will show advantages.

7.4 Still more efficient spectrum utilization

The possibility exists of transmission in even less occupied spectrum than for QPSK, and hence yet narrower radio channel widths (or alternatively of transmitting higher data rates in the same channel). This is possible through **multi-level** data transmission. We now return to the baseband signal, and for simplicity will consider it in bipolar form.

If during each clock interval a single binary digit is to be transmitted, switching between positive and negative, it is only necessary to determine the polarity of the signal. However, it would be possible to assign two distinguishable, possible, positive signal amplitudes (say, one three times the other) and two negative ones. Thus for each clock interval there would now be four distinct states corresponding to +v, +0.33v, −0.33v, v which can thus signal two binary digits, double the transmission rate for the simple binary form, although it is obvious that because the states to be discriminated are now closer together, the noise power must be reduced if the same error rate is to be achieved.

If now this waveform were to replace the binary modulating waveform in BPSK, in signal space the diagram becomes (in the absence of noise) four points on the vertical axis. As usual these spread to patches when noise is added in. This system of modulation would be described as **4-level AM**. For the same maximum signal power the difference between the states is reduced by a factor of 3 (in voltage), so in order to maintain the same error rate the noise must be scaled down similarly, or in other words the S/N power ratio must improve by 9.5 dB (×9). Thus for a 10^{-4} error rate the required S/N ratio is 17.9 dB. If, instead of utilizing the higher data transmission rate we choose to reduce the system bandwidth by a factor of two compared with BPSK (and hence the noise power by the same amount) we still need to increase the received signal power by $9.5 - 3 = 6.5$ dB compared with the binary case, to a value of 14.9 dB. This is much less favourable than QPSK. Data modulation strategies which use one only of *either* the angle *or* amplitude domains can never be as effective as those, like QPSK, using both.

However, if we combine QPSK with 4-amplitude signalling in each channel, the result is signalling four times faster, for an S/N ratio penalty of $9.5 + 3 = 12.5$ dB, requiring an S/N ratio of $8.4 + 12.5 = 20.9$ dB for an error rate of 10^{-4}. This is well within the capabilities of many radio channels, and gives 16 distinct states: **16-level QAM** (or QPSK) Fig. 7.7. A signalling rate of 4 b/s per hertz of bandwidth is available, so that a communications-quality voice channel 3 kHz wide would be able to carry 12 kb/s, while an 8 MHz digital television complex could carry 32 Mb/s.

Obviously, this multi-level signalling technique could be carried further, using eight amplitudes to give three bits per symbol, 16 to give four bits per symbol, and so on. Each increase in the number of

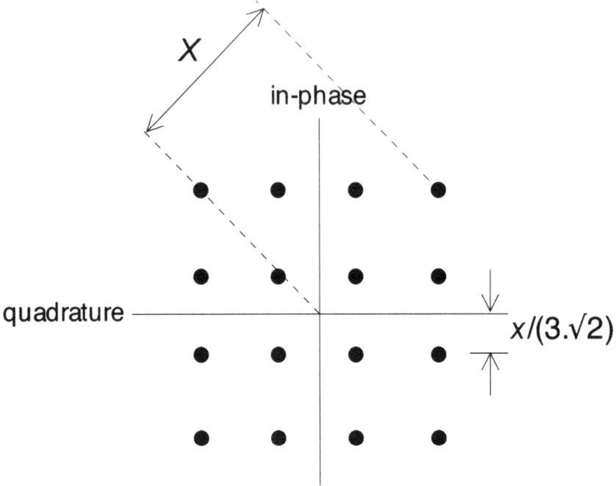

Fig. 7.7
16-level QPSK represented in signal space.

amplitudes, however, will also require an improvement in the S/N ratio for a given error rate. Whereas for four amplitudes the spacing between them is reduced three-fold, requiring a 9.5 dB improvement in S/N ratio, for 16 amplitudes the spacing is reduced 15-fold, so the S/N ratio penalty becomes 23.2 dB, with the result that for the same 10^{-4} error rate an S/N ratio of 31.6 dB is needed at the receiver. This is a high but not impossible S/N ratio for a radio channel, and would permit signalling at a rate of 24 kb/s in a 3 kHz bandwidth using **256-level QAM** modulation, or similarly 64 Mb/s in an 8 MHz complex.

Using sophisticated modem designs it is always possible to achieve high data transmission rates, but the penalty is that the S/N ratio must be increased. This sets a limit to the signalling rates which are possible, and in practice more than 16 distinct amplitudes are not encountered in radio systems. It is possible to make modems **adaptive**; the S/N ratio is continuously monitored and the coding or modulation techniques are adjusted to give the highest data rate (compatible with an acceptable error rate) for the S/N ratio detected at any particular time. Even so, most effective spectrum utilization is achieved by using the whole of the available S/N ratio, and curtailing the bandwidth occupied to whatever is then necessary for the required signalling rate.

7.5 Other modem strategies

All the data modulation schemes considered so far assume synchronous demodulation of the received signal. This always gives optimal performance, but may sometimes be difficult to implement, particularly when frequency offsets are present. For example, if data is being received in a vehicle travelling at speed there will be a doppler shift in received frequency which is velocity dependent. A supersonic airliner travelling at Mach 2.5 and receiving radio transmissions from a fixed station behind it at 150 MHz would find the signals reduced in frequency by as much as 375 Hz, so the effect is significant. Using digital signal processing to extract suitable Fourier components from the received signals it is possible to phase lock a local clock to the signal even so, tracking the variation. The hardware/software sophistication involved, although significant, does not attract much cost penalty. In the past, when hardware costs were much higher, other solutions to this problem had to be found.

The simplest tactic was to transmit data by full-carrier AM and extract it at the receiver by simple envelope demodulation. The disadvantages of this approach are obvious and have already been described for analogue transmission: much of the radiated power is carrier, required only to facilitate easy demodulation, and also the bandwidth is doubled, corresponding to only one bit per hertz. Compared even with BPSK this gives rise to a nearly 6 dB S/N ratio penalty for the same bandwidth, or 9 dB for the same signalling rate (twice the bandwidth). Vestigial sideband transmission (VSB) does better than full AM; the bandwidth is not much more than half for the same signalling speed. However, in both cases it is necessary to make the receiver bandwidth wider than the received signal spectrum, to allow for the frequency shift, which further worsens matters, generally to the extent of a few decibels. Both of these techniques must now be regarded as obsolete.

In the past frequency modulation was also sometimes used for data transmission, and if the bandwidth of the receiver is wide enough this solution can cope with frequency drift. It is not difficult, however, to show that all the problems that make FM uneconomic in spectrum use for analogue transmission apply with equal force to data transmission. The characteristics of the system are almost as unattractive as full AM, with a 6.5 dB S/N penalty relative to BPSK. Although existing FM data systems may still be encountered, the design of new radio equipment using this approach would be unethical in the present era of spectrum congestion, except perhaps for short-range links in the millimetre wave absorption bands, where the risk of interference with others does not arise.

Of far more interest is **differential phase** transmission. Whereas in BPSK each digit is transmitted as the absolute phase (relative to a reference phase) of the transmitted signal, it is a perfectly straightforward matter to encode the signal differentially. Thus, if a **1** is to be transmitted then a phase switch does not take place at the beginning of the clock period, but if a **0** is to be transmitted then a phase switch occurs, the signal amplitude passing through zero as this happens. At the receiver it is then only necessary to multiply the received sequence by itself delayed by one clock cycle. Where no phase difference is found between the signal and its delayed version the product is positive, interpreted as a **1**, otherwise it is negative, interpreted as a **0**. Implementing differential transmission is simplicity itself; all that is needed is to use the original baseband digit stream to form another which transmits the same digit as the previous one (that is, **1** if the previous digit was **1**, **0** if it was **0**) for a **1** in the original, but changes to the opposite (**1** after **0**, or **0** after **1**) for a **0** in the original.

The obvious disadvantage with a differential system is that the error rate must be higher. If the received signal is corrupted in a certain clock period this can produce an error twice, since it will affect both the delayed and undelayed versions. Superficially an approximately doubled error rate looks serious, but in practice it is not. The error rate is such a sharp function of S/N ratio (for example, in the case of BPSK falling from 10^{-4} to 10^{-8} for under 4 dB improvement in S/N) that less than 1 dB increase in S/N will overcome the increased error rate due to differential working.

Questions

1. Why is it unsatisfactory to attempt to constrain the spectrum of a data stream by filtering alone? Consider both high and low frequency components in your reply.

2. An LM radio link in a 5 kHz channel transmits signals with baseband spectrum constrained to the range 200 Hz to 3.3 kHz. Transmitter power is chosen so that an S/N ratio of 26 dB is always exceeded at the receiver, and a raw BER of 10^{-4} can be tolerated. What data rate can be transmitted using multi-level QAM? (18.6 kb/s with 64-level QAM at 25.3 dB S/N)

3. What effects do available power constraints, as in battery-powered portable equipment, have on the design of data systems?

CHAPTER 8

FIGHTING INTERFERENCE: SPACE, TIME AND FORMAT

In a congested radio band it is not only spectrum pollution which can impair the transmission. Radio energy legitimately and necessarily emitted as part of a licensed transmission may interfere with other transmissions, causing their reception to be corrupted. In fact in present-day conditions in very many radio bands, from MF to SHF, this is the most likely type of signal impairment.

Interference between radio signals can be minimized in three basic ways: by exploiting space, time, or format. We review each of these briefly, to define their meaning, and then subsequently consider them in detail.

8.1 Isolation by space

Early radio receivers had very little ability to respond selectively to the signals arriving from their antennas, but interference was not a problem at first because there were so few transmitters in use that receivers were unlikely to be within working range from more than one. Of course, this soon changed. Essentially, however, early users were exploiting their spatial separation to minimize interference.

As we have seen, all radio transmissions fall in power density as the distance from the transmitter increases; they are usually governed by an inverse square or fourth-power law relating received signal to range. Thus if the distance to the transmitter it is desired to receive is less (perhaps very much less) than the distance to an interfering transmitter the effects of interference can be reduced to an acceptably low level. Detailed working out of the use of space separation for this purpose depends on the actual mathematical law relating signal strength to distance. A number of important cases of this kind will be considered in what follows.

Isolation by spatial separation leads naturally to the notion of frequency assignment plans. Radio regulatory authorities and others responsible for frequency assignment try to minimize the potential for interference by suitable spatial patterns of assignment which maximize the distance between those sharing a common frequency. Implicit in this idea is that transmissions on different frequencies will not interfere – an example of isolation by format, which we will now consider.

8.2 Isolation by format

Another way to separate radio transmissions is by radiating them in mutually orthogonal formats. The most familiar of these is to separate them in the **frequency domain**, by modulating baseband signals on to carriers at different frequencies. Provided that the spectrum for each transmission does not overlap significantly with the others, they can be effectively separated by the use of **filters**, usually band-pass. This technique goes back to the very beginning of radio technology, and is the origin of the radio tuning dial, which sets the centre frequency of the internal band-pass filtering function.

In the last quarter century, however, it has become widely appreciated that frequency is just a special case of **sequency**, and that consequently it is possible to generalize frequency orthogonality in a new and powerful way. Frequency is conventionally defined as half the inverse of the time interval between two successive zero-crossings of the carrier wave, and it is assumed that these occur at regular intervals. (The fact that the carrier is usually sinusoidal is only a matter of convenience, as is the choice of the inverse rather than the direct value of the time interval.) Frequency is thus a one-dimensional variable.

However, if we consider a wave sequence (for simplicity they may be taken as rectangular waves) in which the interval between zero-crossings is not always the same but varies, we can specify the wave by an ensemble of numbers, for example the inverse of the time interval between the first and second zero-crossings, between the second and third, the third and fourth, and so on. In the case of a truly random sequence it is impossible fully to specify it, because an infinite set of numbers would be required, but if the sequence repeats after a certain time the set of numbers required is finite, though it may be large. This set of numbers defines the sequency of the wave, and it may be thought of as a space vector. Unlike frequency, it is no longer one dimensional but many dimensional, thus defining a point in a **hyperspace**. Each such point defines a distinct wave sequence which could act as a carrier for a signal. We can build filters to extract signals which are not one dimensional in the sequency space, as frequency filters do, but are able to filter hyperdimensional sequencies.

Since all real filters are of finite bandwidth, going from one-dimensional frequency to many dimensional sequency greatly enhances the number of distinct carriers which can be separated. The passband of a frequency filter corresponds to an interval on the frequency scale, whereas the passband of a sequency filter corresponds to a volume around the point in hyperspace which defines the central sequency. For historical reasons, sequencies of dimensionality greater than one (that is, excluding frequencies) are usually described as **codes** and orthogonality between signals is obtained by using **orthogonal codes**. Generalizing frequency into sequency opens the door to powerful orthogonal format techniques, made practicable by the effectiveness and low cost of modern digital signal processing hardware.

8.3 Isolation in time

In the early history of radio, soon after its introduction, a day inevitably came when there were enough transmitters in use for mutual interference to begin to cause problems. One solution sometimes adopted was for the parties to agree to transmit at different times, and even today licences to transmit occasionally have time-of-day limitations to reduce interference potential with other users. Obviously, transmissions which take place in different time intervals cannot normally interfere with each other.

Isolation in time, either to facilitate multiplexing or multiple access systems, remains in widespread use, and is, for example, the basis of the GSM cellular radiotelephone system now adopted worldwide.

This, however, does not exhaust the significance of the time domain in radio engineering. Many patterns of radio use require the frequent transmission of relatively short messages at irregular time intervals, which may be long relative to the message. If only a single channel is available for these transmissions there will sometimes be **contention** between two or more transmissions originating close in time, which must be suitably resolved. Alternatively time-limited messages from multiple sources may be carried on a group of channels allotted for that purpose, users being assigned to a unique channel for one message only, and possibly to another next time they wish to communicate. This is known as **trunking**, and is obviously a more economical use of the spectrum than permanently assigning a channel to each potential user.

8.4 Protection ratios

All of these methods of separating wanted from interfering or polluting signals aim to reduce the amplitude of the latter relative to the former in the signal reaching the demodulator – ideally we would like the interfering signal amplitude to be reduced to zero. The ideal case can rarely be met, however, and it is usually necessary to accept a certain ratio of wanted to interfering signal amplitudes. Provided that this is sufficiently large, the effects of the interference may usually be reduced to negligible proportions. The minimum acceptable ratio of wanted to unwanted signal powers at the input to the receiver, which just reduces the probability of received signal degradation to a specified low level, is known as the **protection ratio**, and is a crucial parameter in the design of radio systems. It varies widely depending on the circumstances.

When two signals are received together the consequences of interference depend on the technical characteristics of both the wanted and interfering transmissions, which naturally need not be the same. Even so, it is possible for each particular combination of wanted and interfering signal types to define a protection ratio, the minimum ratio (normally expressed in decibels) of wanted to unwanted signal power at which just acceptable reception of the wanted signal is obtained. But how is 'acceptable' defined in each such case? It can only be determined according to some user's criterion of permissible received signal degradation, which varies according to the particular radio service being considered.

For example, in a digital system raw **bit error rate** (**BER**) is most commonly adopted as the measure of the quality of the radio channel. BER is the proportion of bits in a random binary stream which are incorrectly received, that is with '0' being substituted for '1' or conversely. However, this does not completely resolve the problem, because the BER which can be tolerated depends on the sensitivity to error of the information to be transmitted. In digitally encoded voice or music, an isolated error may do no worse than cause a slight extraneous noise. By contrast, if the digits to be transmitted specify a banking cash transfer the tolerance of error will necessarily be much less. The range of tolerable error, from maximum acceptable to minimum, is thus wide, but in practice only a very few digital radio systems are designed to have raw bit error rates worse than 10^{-3} (1 error on average in every 10^3 bits transmitted), and most require 10^{-4} or better. With some cost to transmission rate, **error detection and correction codes** can then be used to reduce the final system error rates to an acceptable lower level, in accordance with user requirements. If the digital transmission technique to be used is known for both interferer and wanted signal, protection ratios can be calculated objectively, at least in principle.

So, in practice how much isolation is required between wanted and unwanted signals, in order that the former may be received satisfactorily? Typical protection ratios for some of the commoner examples of the wide variety of possible combinations follow; they assume that the interfering signal is similar to the wanted signal. (We can also regard the minimum acceptable S/N ratio in a system as equivalent to the protection ratio when the 'interfering signal' is in the form of noise, so S/N ratio issues need not be considered separately.) For digital transmissions, the required protection ratio depends on the type of modulation, a robust binary system like BPSK being able to tolerate more interference than a multi-level system (where the 'distance' between the levels, in phase or amplitude, is less). However, as already explained, more robust systems are characterized by lower data transmission rates. We have also noted that for each modulation system there exists a curve relating bit error rate to S/N power ratio, and although the results are quite seriously modified if noise is replaced by coherent interference, they remain a guide to the likely outcome. Practical systems using 2- to 16-level QPSK modulation may have protection ratios in the range 9–24 dB for error rates of 10^{-4}, depending on the nature of the interference. Systems designed for higher data rates or lower error rates will require higher protection ratios. These are determined empirically with knowledge of the likely type of interference.

8.5 Protection ratio in analogue systems

Although the move to digital technology is gathering pace, it remains true that the majority of current radio systems use analogue transmission, and will continue to do so for many years to come. Here the criteria used to set the protection ratio are quite different.

In the case of analogue broadcasting, a subjective test of perceived quality is often used (ITU 1997a). For both sound and vision broadcasting, the protection ratio is defined as the signal-to-interference power ratio at which the limiting subjectively assessed degradation is reached. In the case of sound broadcasting, panels of listeners score what they hear over the channels on a scale of acceptability. Because individuals vary greatly and are not consistent in time, typically a panel of twenty or more listeners will participate in tests carried out over an extended period. With television much the same procedures have to be followed (ITU 1997b). This is a slow and expensive business, but scientific understanding of psychoacoustics is not yet good enough for these matters to be determined objectively, so subjective testing cannot be avoided at present.

For analogue voice and music transmissions the position is complicated because the protection ratio depends on the mode of modulation adopted. Due to the low average modulation index of analogue speech (as a result of the high peak-to-mean ratio of the speech waveform and frequent silences), most (typically over 90%) of the energy radiated by AM or FM transmitters is in the form of a component at the carrier frequency, and the main audible disturbance due to interference is from carrier interactions, heard as heterodyne whistles or beat notes. Carrier interactions remain subjectively unacceptable until the difference between the two signals is large, particularly so because they are heard as continuous and clearly perceived in transmission silences. Depending on the acceptable quality of audible signal, the protection ratio may be from about 17 dB (lowest FM) to 40 dB or higher (high quality AM) (ITU 1997c, 1998).

Something similar applies to analogue television broadcasting. With the vestigial sideband (VSB) modulation now universally adopted, when subject to interference analogue television typically shows moiré patterns in the picture, defective colour, and in severe cases loss of synchronization. Here too, aside from catastrophic interference, objective calculations are unsatisfactory and subjective judgements of moderate impairment of the picture quality are required. Satisfactory service is generally found to require protection ratios of 50–60 dB. These are high, compared with sound broadcasting, because the eye is very sensitive to small visual disturbances of the picture (ITU 1997d).

It is intelligibility, in marked contrast, which is the primary concern for non-broadcast analogue speech channels, ranging from taxi radios through police, fire and ambulance services, to air traffic control. This may be measured by reading lists of random words over the radio channel and scoring what proportion is correctly identified. Again, the protection ratio is defined as the signal-to-interference power ratio at which the intelligibility falls to the minimum acceptable level. Sometimes the **SINAD** ratio (signal-to-noise + distortion) of the demodulated signal is specified, corresponding to this lower intelligibility limit, in which case the protection ratio is defined when this SINAD is just achieved. In practice the protection ratio depends here too on the mode of modulation adopted. Three kinds are commonplace: amplitude modulation (AM), angle modulation (FM and PM) and linear modulation (LM), or occasionally SSB (single sideband). Carrier interactions are again the main cause of difficulty, and for AM and narrowband FM (**NBFM**) protection ratios commonly considered acceptable for communications-quality services range from about 14 to 24 dB.

Analogue LM has no carrier, so in the absence of a pilot tone interference effects would simply be a consequence of sideband interactions. These are far better tolerated by listeners because they are transient and not perceptible in silences except as a faint, distorted second voice, which hearers are psychologically well adapted to ignore. Pilot tone interactions are subjectively similar to the carrier interactions with AM or FM, producing a continuous tone, but are much weaker. So, if the pilot tone were at −13 dB relative to the AM carrier (and in practice this would be considered a high figure) the interaction between two would be reduced by 26 dB, and is consequently much less audible. Protection ratios are thus lower, typically 10–15 dB for good intelligibility (depending somewhat on pilot tone level).

Although intelligibility alone has dominated much past thinking about radiotelephone systems, it is now recognized that it is necessary but insufficient. Users greatly value speaker recognition also, and systems which offer it achieve a market advantage. Even military users evaluate speaker identification positively, as a means of authentication. These requirements often push up minimum acceptable protection ratios.

In any event, the necessary protection ratios can only be achieved by exploiting the space, format and time parameters of the system. It is to these issues that we now turn.

Questions

1. In the case of analogue television signals, how would the effect of pollution in the form of passband noise compare with a single frequency? (the former causes short-range picture intensity fluctuations, the latter bars or moiré patterns, with possible loss of synchronization)

2. Why are protection ratios for digital systems often lower than those for analogue systems?

3. In the case of multi-level QAM, how would you expect the protection ratio for a given raw BER to vary with number of levels?

CHAPTER 9

EXPLOITING THE SPACE DIMENSION

All radio transmissions consist of large numbers of quanta of radio energy, particles moving with the speed of light. In all but the very rarest of situations these streams of quanta are diverging. In consequence, as we have seen, the density of quanta falls with distance from the transmitting location, and hence fewer quanta will be captured by a receiving antenna of fixed aperture at a more distant site. This is the familiar picture that in all radio systems the received signal power falls with range. In addition, quanta may actually be absorbed, by the Earth in the case of surface propagation and increasingly by the atmosphere above a transmission frequency of 20 GHz.

Thus provided that the distance from the receiving location to the interfering transmitter is much larger than the distance to the wanted transmitter, it may be possible to maintain a sufficiently large ratio between the relative received signal strengths for the effects of the interference to be negligible.

9.1 The basic arithmetic

Effective isolation from interfering radio transmitters by ensuring that they are far enough away depends critically on the law relating signal strength to distance from the transmitter. For the case of transmissions in space, the signal power density falls as an inverse square law, simply because the total area irradiated increases as the square of distance. In the case of VHF or UHF scattering

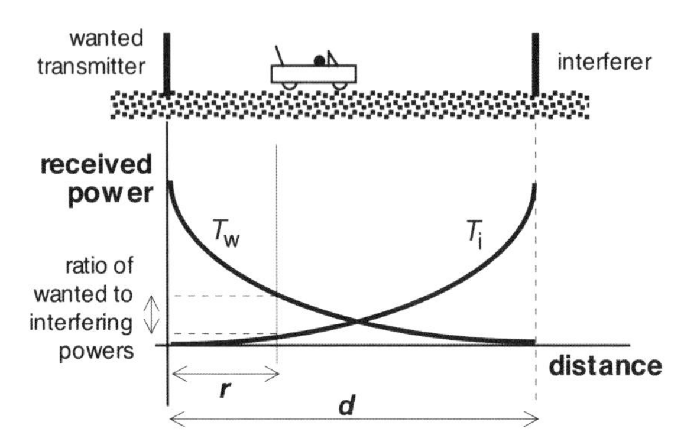

Fig. 9.1
Variation of received signal power with distance from the wanted and interfering transmitters.

transmission near the ground, with no line-of-sight path, the mean received power varies as an inverse fourth power of the range, but there is considerable signal fluctuation about this mean, the Rayleigh fading. Under thickly wooded wet trees, in the VHF and UHF bands the mean signal may fall as rapidly as the inverse sixth power of distance from the transmitter. It therefore seems a good starting point to consider the simple case of signals with an inverse nth power law of variation with range, deriving from omnidirectional transmitting antennas (so that the angle of propagation is unimportant).

What service range can we expect from a transmitter when there is another not too distant, capable of causing radio interference? A transmitter T_W emits the signal that is to be received at a point distant r, and simultaneously at a distance d (assumed on the same line as the receiving station, to keep the geometry simple), a second transmitter T_i radiates signals of equal power in the same format, which interfere with the wanted signal at the receiving site distant $(d - r)$ from it (Fig. 9.1).

At the receiver the ratio of wanted to interfering signal powers, assuming that propagation is determined by an inverse nth law, is:

$$q = \left(\frac{d - r}{r}\right)^n \qquad (9.1)$$

If this ratio is greater than some minimum value, say $q_{\min}$ (to be evaluated), the wanted signal will be received without unacceptable impairment due to interference. So for satisfactory reception:

$$\frac{d-r}{r} \geq (q_{\min})^{1/n} \tag{9.2}$$

But this gives the maximum service range of the wanted transmitter, r_s, where:

$$r_s = \frac{d}{1 + (q_{\min})^{1/n}} \tag{9.3}$$

and from this the **interference range** r_i of the transmitter T_i can also be obtained as:

$$r_i = d - r_s = \frac{d \cdot (q_{\min})^{1/n}}{1 + (q_{\min})^{1/n}} \tag{9.4}$$

However, when frequencies are being assigned to users by the radio regulatory authority, the location and service range of existing transmitters is known, and what is of interest is at what distance from the existing transmitter the same frequency assignment can be given to a newly installed transmitter (carrying similar modulation) without unacceptable interference. This distance is known as the **frequency reuse distance**. Evidently its value is:

$$d_r = r_s\{(q_{\min})^{1/n} + 1\} \tag{9.5}$$

We can therefore use these expressions to calculate the maximum service range where the transmitter locations are known, or to arrive at the frequency reuse distance. Note also that the interference range of a transmitter can be written:

$$r_i = r_s \cdot (q_{\min})^{1/n} \tag{9.6}$$

9.2 Minimum ratio of wanted to interfering signals

Obviously in either of the cases considered q_{min} is of the utmost importance and we now investigate its value. The minimum signal-to-interference power ratio q_{min} is the product of three factors:

$$q_{min} = q_1 \cdot q_2 \cdot q_3 \tag{9.7}$$

The first of these arises from the **protection ratio** required. As stated, this is the ratio of wanted signal power to interference power which will just give a satisfactory, received signal after demodulation. For digital transmissions, we have seen that 'satisfactory' here means the ratio which will just give the required maximum permissible raw error rate, whilst for analogue transmissions it means the lowest ratio which gives a result rated satisfactory in subjective testing. (In the case of digitized sound or vision transmissions the acceptable maximum error rate is also determined by subjective testing.)

If the protection ratio in decibels is R_p, then:

$$R_p = 10 \log_{10}(q_1) \tag{9.8}$$

from which q_1 may be obtained.

The second factor, q_2, is a **safety margin** which may in some cases be included to allow for signal strength fluctuation. A factor like this will be included whenever the propagation characteristics are such that there is a differential signal variation between the wanted and interfering signal. It can be omitted where either there is no signal fluctuation of this type, or where it affects both wanted signal and interference similarly.

For example, whereas in free-space propagation the signal strength may be calculated with some precision, in the case of urban UHF and VHF propagation, which is entirely by scattering, only the mean value of the signal can be approximated and the actual value at a particular location is statistically distributed about this mean value calculated for that local area. Over variations in position of up to a hundred wavelengths the signal is characterized by a Rayleigh statistical distribution of amplitude and random phase about a constant mean value. Longer-range signal variation may be characterized as an inverse fourth power variation of the mean of the statistical distribution. This is the key to the satisfactory description of scattered radio transmissions.

Since, in the scattering case, the calculation of the signal is probabilistic, not deterministic, it is necessary to decide on the grade of service which is required. What probability can be accepted that the wanted signal will not exceed the interference by the desired margin, and the system will therefore malfunction? Obviously this depends on user requirements, but for many users a 5% probability of malfunction might be unacceptable, 1% would be acceptable, and significantly better than that would be thought a superior service. For 1% malfunction, as we have seen, an 18.5 dB margin is required for a 99% surety of service (1% failure). (In practice 20 dB is a commonly used 'round figure' safety margin, which should give a little lower failure rate.) Supposing a 1% malfunction probability is acceptable, this would suggest that q_2 should be set at 18.5 dB ($\times$ 70), and if either a higher, or lower grade of service were required this figure would be adjusted accordingly.

So much for the scattering environment. In the case where signal variation occurs due to atmospheric absorption, important particularly for terrestrial radio links in the millimetre wavebands, the situation is a little different. Signal absorption is principally dependent on the amount of water vapour in the air, being very severe in rain. However, it is also proportional to path length, and therefore the additional absorption will always be worse for the interfering than for the wanted signal. Thus if the system is designed for dry air conditions it will always be better when there is rain or mist, so a q_2 term need not be included in this case.

As for q_3, this takes account of the ratio of the powers of the interfering and wanted transmitters. If we generalize the analysis to the case where the transmitter powers are no longer equal, equation (9.1) becomes:

$$q = \left(\frac{d-r}{r}\right)^n \cdot \frac{P_\mathrm{w}}{P_\mathrm{i}} \tag{9.9}$$

where P_w is the wanted transmitter power and P_i the interferer. Hence, moving the power ratio to the LHS of the equation, q_{min} is given by:

$$q_{min} = q_1 \cdot q_2 \cdot \left(\frac{P_i}{P_w} \right) = q_1 \cdot q_2 \cdot q_3 \tag{9.10}$$

We have assumed that both have omnidirectional antennas, but this limitation can be removed by taking q_3 as the ratio not of the actual transmitter powers but of their EIRPs (equivalent isotropic radiated powers). As explained above (Chapter 2), the EIRP is the actual transmitter power (as delivered to the antenna) multiplied by the antenna power gain *in the direction of the receiver*, obtained from the antenna polar diagram. So:

$$q_3 = \left[\frac{P_i}{P_w} \right]_{EIRP} \tag{9.11}$$

9.3 Space isolation examples: 1. Free space

The case of free-space propagation is the simplest. There are genuine space uses, for example transmissions between low Earth orbit (LEO) satellites. At frequencies low enough for atmospheric absorption not to be significant (say below 20 GHz) remote-from-Earth terrestrial examples can approximate well to the free-space condition, like aviation radio, and even propagation between antennas on masts high enough to avoid terrain reflections. The free-space case therefore is evidently of real importance.

In free space propagation is subject to an inverse square law, so $n = 2$. What are the q values? In this case there is no reflection or scattering and the signal is not subject to statistical variation, so the q_2 term is not required. For the purposes of this example, we will also assume that the EIRPs of the wanted and interfering transmitters are the same. Thus only the protection ratio is left, so:

$$q_{min} = q_1$$

As already indicated, its value may typically range from under 10 dB (×10) to over 50 dB (×100 000), depending on the type of transmissions and required grade of service.

We can now calculate the interference range and frequency reuse distances. Absolute values are less interesting for this discussion than normalized values, that is expressed as a multiplier of the service range. The normalized interference range is:

$$R_i = \frac{r_i}{r_s} = \sqrt{q_{\min}} \tag{9.12}$$

Taking the extremes of protection ratio, the interference range is from just over three to just over three hundred times the service range. In this latter, admittedly unrealistic, case, the transmitter (in true free-space conditions) causes interference at more than three hundred times the range that it can provide a satisfactory service. (This ratio would not be seen in practice in near-to-Earth situations, since over ranges as large as this the assumption that propagation is at all points sufficiently remote from the Earth's surface is unlikely to be valid, the curvature of the surface no longer being negligible. However, it could happen in deep space.)

This latter ratio of interference to service range is a disastrous result, since it means that to provide services in contiguous areas very many different frequencies will be required if users within this very large interference range are not to suffer degradation of service. Obviously, with a more modest protection ratio things are not quite so bad, but it is clear that to provide satisfactory services under free-space conditions requires very many frequencies. Even a 30 dB protection ratio would be impossible to accommodate, in practical terms.

What can be done about this? A number of conclusions emerge irresistibly. The first is that in free-space propagation (or anything that approximates to it) it is important to choose methods of modulation with the lowest possible protection ratio, and this inevitably drives the system designer toward digital solutions. For example, the protection ratio required for 10^{-4} BER with BPSK has been shown to be 8.4 dB, so in this case the interference range is only 2.66 times the service range, which is far more manageable. QPSK needs a 3 dB better protection ratio but, for the same data transmission rate, occupies a channel only half as wide, so in terms of total spectrum occupancy it is similarly advantageous.

The other obvious conclusion about free-space propagation is that if the system makes it at all possible, the interference range can be much reduced if instead of q_3 being unity as assumed so far (that is, the EIRPs of the wanted and interfering transmitters are the same) a value much less than one can be arranged. This may often be possible by using directional antennas, with nulls or very low gain figures in the interfering direction but main lobe maxima in the directions in which service is required. How practical this may be depends on the geometry of the system. In the case of point–point terrestrial radio links, with antennas mounted at a sufficient elevation to approximate free-space propagation, highly directive antennas will anyway be preferred to reduce the required transmitter power, and in this case q_3 is likely to have a large negative decibel value. By contrast, terrestrial television broadcasting will often achieve near free-space propagation when received by rooftop antennas but transmission is necessarily omnidirectional in the horizontal plane. This, along with the high protection ratio required for analogue television, results in very large interference ranges, limited largely by terrain features including the Earth's surface curvature, so that an inverse square propagation law no longer holds.

9.4 Space isolation examples: 2. The scattering environment

The propagation of radio waves by scattering is the dominant mode in the VHF and UHF bands provided that the antenna elevation is low compared with terrain features such as buildings, hills and perhaps even vehicles. With rare exceptions, it applies also to use of these bands within buildings. It results, as we have seen, in a received signal strength which is statistically distributed in amplitude, phase and time of arrival. The mean value of the signal varies as the inverse fourth power of range, so $n = 4$.

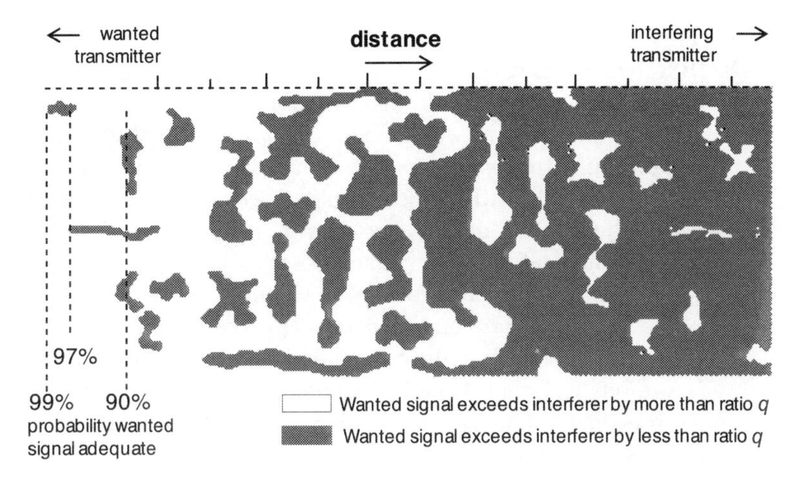

Fig. 9.2
In a scattering environment the local variation of signal strength near the edge of the service area is a complex 'froth'.

For a given radio system the value of the protection ratio is unaltered by the mode of propagation, and gives q_1 as before, but now a q_2 term is needed to guarantee a required grade of service against the consequences of Rayleigh fading, as already described (Fig. 9.2). This is unlikely to be less than 20 dB (giving an approximately 1% probability of signal falling below the desired minimum), and might be 30 dB for an approximately 0.1% drop-out probability, or could be some even higher value for still better security of service, all calculated from the cumulative Rayleigh distribution.

Beginning, once more, with the case where the wanted and interfering EIRPs are equal, so that q_3 is unity, the value of $q_{min} = q_1 \cdot q_2$ (expressed in decibels) may range from under 30 dB for a digital system with 1% tolerated failure probability, to some 80 dB for a high quality analogue system with comparable rate, up to some very high (and quite impracticable) values for hypothetical analogue systems of very high reliability. Again we can use these values to calculate service and interference ranges, which are more favourable in this case due to the inverse fourth-power law governing propagation.

Taking the lowest value of q_{min} as 30 dB (10 dB protection ratio plus 20 dB for statistical variation), the interference range is just over 5.6 times the service range. Although because of the Rayleigh fading 'safety margin', this is worse than the 3.2 times of the 10 dB protection ratio free-space case, it is not seriously so. Taking a protection ratio of 30 dB (which would cover many analogue uses and most high-speed data systems) with the usual 20 dB safety margin, the normalized interference range is just under 18, compared with over 31 in the free-space example considered above. Here too the situation can often be improved where directional antennas can be used to reduce the interfering EIRPs in significant ways, although of course there are some systems where the use of omnidirectional antennas is unavoidable, as with radiotelephones.

Note that the advantage given by a lower protection ratio is much less in the scattering case. It is also clear that in the great majority of cases it is more (and possibly much more) spectrum conserving to work with scattering than with free-space propagation. This is fortunate, since in the majority of terrestrial applications of radio we have no option but to work with scattered signals anyway.

9.5 Frequency assignment plans

Suppose that many transmitters operate in different locations but within potential interference range of each other. Obviously, those sites sufficiently close to be able to interfere with each other's signals cannot be allowed to operate without some additional means to avoid degradation of each other's services. They could be arranged to operate at different times, for example, or with different, non-interfering signal formats. The commonest way to avoid their mutual interference is to assign them to different channels in the band. Separation by time and format will be considered further in the following chapters; for the moment we will assume that it is enough to assign the transmissions to different carrier frequencies. At the same time the number of channels available is strictly limited, so it must be an objective to minimize the number in use. Frequencies should therefore be assigned in a pattern designed to avoid the interference risk, whilst minimizing the number of different channels occupied. How shall we go about developing the frequency use pattern to achieve this end? We begin by considering frequency assignment in a service area which may be approximated by a flat rough plane.

Conventionally the flat plane is '**tiled**' with contiguous regular hexagons each representing the service area of an individual transmitter, and to each tile, or **cell**, a suitable frequency is assigned so as to minimize the overall interference potential. Hexagons are chosen because the shape approximates fairly well to the near-circular coverage area of a real omnidirectional transmitter on flat ground, yet hexagons will fit together without spaces between them. (Sometimes other tiling shapes are used, particularly squares, but this leads to a more approximate treatment.) We begin with the simplest case, in which transmitters are located at the centre of cells and all antennas, transmitting and receiving, are omnidirectional. Frequencies must be assigned so that the same one is not used in any two cells close enough for interference to result.

The simplest pattern uses only three frequencies (A, B, C) (Fig. 9.3). Each A cell is surrounded by a ring of B and C cells, each B cell is surrounded by A and C cells and each C by A cells and B cells. The distance from the centre of a cell (where the transmitter is normally located) to the nearest point of a cell on the same frequency will be seen to be $2H$, where H is the cell side. The frequency reuse

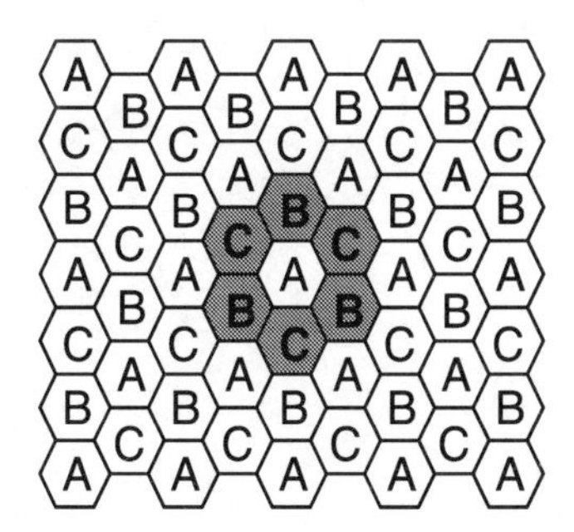

Fig. 9.3
A 3-frequency 'tiling pattern', in which one B–C–B–C circle surrounds each A (example shaded), and B and C are similarly each surrounded by a guard circle.

distance, from the centre of a cell to the centre of the nearest cell able to use the same frequency, is $3H$. Assuming scattering propagation, so that the mean received signal varies as the inverse fourth power of range, the mean wanted to interfering signal power ratio (for a receiver at the nearest point to the interfering transmitter, at a hexagon corner) is $\times 16$ or 12 dB. Corresponding to the factor q in the previous analysis, this value is insufficient even for robust digital or LM analogue systems. Thus the simple 3-frequency assignment pattern with omnidirectional transmissions from the centre of cells is of little practical use. There are two alternative ways to improve the situation: the use of multiple transmitters and directional antennas within cells, or the use of more frequency assignments to increase the space separation between co-channel users. We deal first with the latter.

Using four frequencies, the reuse distance increases to $2 \cdot \sqrt{3H} = 3.5H$, and the q factor at the cell edge is 19 dB. Going to nine frequencies gives a frequency reuse distance of $3 \cdot \sqrt{3H} = 5.2H$ and a q factor of 28 dB, which is large enough to be a useful value in system design (Fig. 9.4). As the number of frequencies assigned increases an ever larger mean protection ratio is attained, but the improvement becomes progressively smaller.

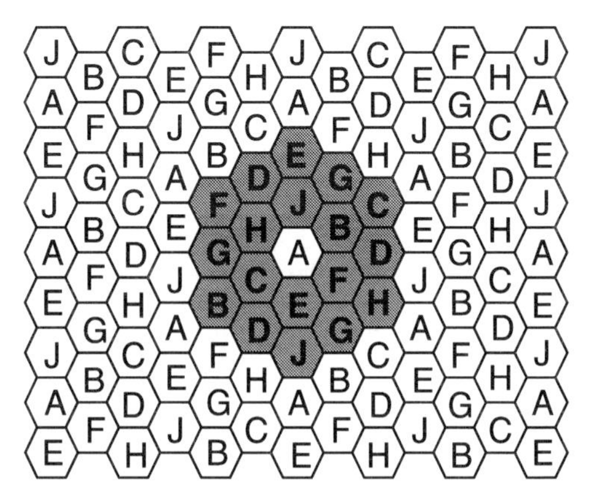

Fig. 9.4
A 9-frequency 'tiling pattern', in which each assignment is surrounded by two guard circles (example shaded).

The normalized frequency reuse distance is related directly to the number of frequencies. Various formulae are given in the literature for the number of frequency assignments, N, required for a tiling pattern; provided that the reuse distance d_r is restricted to integer values of H the following approximation is as good as any:

$$N = \left(\frac{n}{2}\right)^2 + \frac{n}{2} + \frac{5}{8} + \frac{3}{8} \cdot (-1)^n \tag{9.20}$$

where $n = d_r/H$. Some values are calculated in the following table.

Normalized reuse distance n	**Number of assignments** N	**Signal ratio** q **(dB)**
1	1	0
2	3	12
3	4	19
4	7	24
5	9	28
6	13	32

The improvement in protection ratio gets progressively less as the number of frequency assignments is increased, so there is little interest beyond thirteen assignments.

All of this assumes that frequency assignments are made on a flat plane. Although this may be a good approximation in some situations, obviously it does not fit in all. Hills can give rise to major radio shadowing, and thus effectively isolate co-channel transmitters which might otherwise interfere with each other. Mast heights at transmitting sites are obviously critical in this case, and calculation of effective path losses follows a conventional pattern, taking account of diffraction effects where necessary. In general, and with proper attention to frequency assignment procedures, spectrum utilization is generally more efficient in hilly than in flat terrain, due to the compartmentalizing effects of higher ground, combined with the fact that the highest population densities are in the valleys. Software packages exist which enable automatic frequency assignment patterns to be computed from terrain maps,

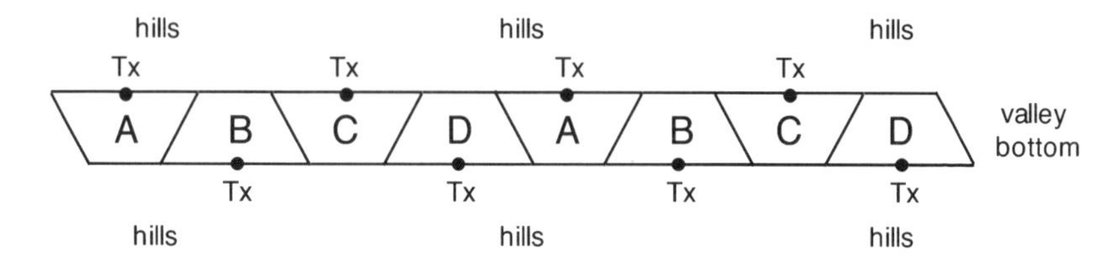

Fig. 9.5
A 4-frequency 'tiling pattern' to give coverage of a valley bottom.

although at present their accuracy is such that the results must be confirmed by field trials.

A case of particular interest is where radio services are required in a long valley which is well defined by a hill ridge on either side. In this case interference propagation is approximately one dimensional, along the valley, and good ratios of wanted signal to interfering signal (q) can be realized with fewer frequency assignments than in the two-dimensional, open-plain case (Fig. 9.5). Directional antennas with their main lobes facing into the valley are positioned on the hill sides at intervals along the length of the valley, maximum advantage being obtained by alternating between the two sides. The transmitting sites can be approximated by half-hexagon coverage areas. For three frequency assignments the signal ratio q is 12 dB as in the two-dimensional case, but the number of assignments rises less rapidly with q, thus for 28 dB N is only 6.

9.6 Minimizing the interference area

Although generally the interference range is more favourable for the inverse fourth-power (scattering) case, clearly for optimum spectrum use the smaller the interference area the better. Is there anything more that can be done? Hitherto the geometry considered is of an omnidirectional transmitter in an ideal flat (but rough, and hence scattering) terrain, providing an approximately circular service area around the transmitting site and a larger concentric interference area. Two techniques are available for improving matters: multiple transmitting sites and directional antennas. We begin with multiple sites.

It is not particularly difficult technically to arrange that a number of transmitters on different sites operate on the same carrier frequencies, that is radiate coherently. This can be done by phase locking carriers to a pilot frequency, which can be distributed to the various sites either by cable, optical fibre, terrestrial radio links or satellite. If all the transmitters are identically modulated with the signal to be transmitted, which again is not particularly difficult to achieve provided the signals are digital, it is intuitively obvious that a mobile receiver can obtain coverage from any of the transmitting sites.

This arrangement, called **synchronous transmission** or **simulcasting**, makes it possible to extend the coverage area beyond what can be achieved with a single transmitter and by judicious siting of the multiple transmitters to cover areas on the ground which are of shapes other than circular – a frequent requirement, needless to say. We shall also show that arrangements of this kind lead to a more favourable ratio of coverage to interference area.

However, before going into detail on such a configuration, a difficulty must first be resolved. Since they are coherent, what will be the effect of interference between the multiple transmitters? Considering the simplest case of two transmitters each radiating an identical transmission, on the line joining the two transmitters how will the signal strength vary? If we were concerned with simple free-space transmission, we would expect to see the signal strength varying cyclically as the receiver moves along the line, the signals from the two transmitters being successively in-phase and in antiphase. Midway between the transmitters, where the two received signals are of equal amplitude, deep signal nulls will occur, separated by half wavelengths, with maxima in between them. This is simply the pattern, familiar from optics, of interference fringes, and reception will fall below acceptable levels in the minima. For this reason multiple transmitter systems are little used in a free-space environment; however, in the case of scattered signals the amplitudes are Rayleigh distributed about a slowly varying mean and all phases are equiprobable, so well-defined interference patterns are not seen. How, therefore, will the two received signals combine in this case?

It is not difficult to show that the sum is also a Rayleigh distributed signal with mean power equal to the sum of the powers of the two received signals. We ignore modulation and consider only radiated carriers. If the receiver has a bandwidth B and the signal from one transmitter after scattering is considered maximally randomized, then for transmitter (1) the received signal is equivalent to a white noise function $e_1(t)$ filtered through a bandpass filter function F of the correct centre frequency for the transmissions and bandwidth B. The output from this filter will obviously have random phase, and it is not difficult to show that the amplitude is Rayleigh distributed. The same will be true for transmitter (2). Thus the received signal is:

$$e_r = F[e_1(t)] + F[e_2(t)] \tag{9.14}$$

where $e_2(t)$ is another white noise function of different power from $e_1(t)$. But both addition and filtering are linear operations, and therefore the order in which they are applied is unimportant (that is, they are commutative). Hence:

$$e_r = F[e_1(t) + e_2(t)] \tag{9.15}$$

But the sum of two white noise functions is also white noise, its mean power equal to the sum of the mean powers of the two. However, when this resultant white noise is filtered it yields the usual Rayleigh distributed signal, of mean power equal to the sum of that of the two received signals. Thus in the scattering propagation environment the use of multiple synchronous transmitters does not give rise to problems of mutual interference, and the mean powers of various transmitters simply add at the point of reception. The interference fringes characteristic of free-space propagation consequently do not occur, and although there are still nulls in the received signal they are only those of the Rayleigh distribution, which can be reduced to an acceptably low probability by an adequate signal strength margin q_2.

As an illustrative example of how synchronous transmission improves spectrum utilization, we begin by considering four

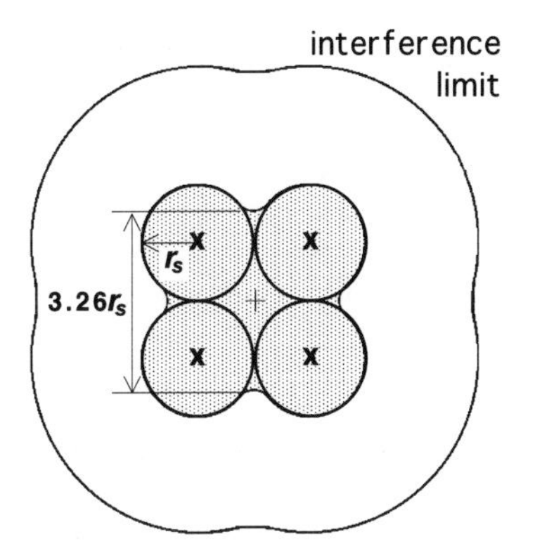

Fig. 9.6
Four synchronous transmitters (each sited at **x**) produce a roughly square service area (shaded) and a smaller interference area than with a single central transmitter.

transmitters in a square configuration on a flat plane (Fig. 9.6). If each transmitter has a service range r_s, then the transmitters need to be sited at the corners of a square of side $2r_s$. The centre of the square is then at a distance $\sqrt{2} \cdot r_s$ from the transmitters, which each therefore deliver one-quarter of the minimum signal power for service at the centre. Since there are four of them the total power is sufficient, and service therefore extends to the centre of the square. By a similar argument, on the two lines which pass at right angles midway between the pairs of transmitters, the service signal strength is exceeded out to $1.63r_s$ from the centre. Thus the area within which service is provided is a little greater than defined by a square of side $3.26r_s$, and the service area is:

$$\begin{aligned} A_s &\geq (3.26r_s)^2 \\ &\geq 10.6r_s^2 \end{aligned} \tag{9.16}$$

The interference radius from each transmitter is r_i and hence the interference area is:

$$A_i \leq 4r_s^2 + 4r_i \cdot r_s + \pi r_i^2 \tag{9.17}$$

So the ratio of interference to service area in this case is:

$$R_4 = 0.38 + 0.38\frac{r_i}{r_s} + 0.3\left(\frac{r_i}{r_s}\right)^2 \tag{9.18}$$

The comparable ratio for a single transmitter is of course:

$$R_1 = \left(\frac{r_i}{r_s}\right)^2 \tag{9.19}$$

For scattering propagation, it has already been shown that R_1 varies from 5.6 upwards. Taking this lowest value of 5.6, R_4 is equal to 2.4, so even in this case the ratio of interference area to service area is more than twice as favourable using the four transmitters than with one. For R_1 equal to 10, R_4 is 4.6 only. In this example four transmitters replace one; the larger the number of transmitters which are used the larger the improvement in the interference area to service area ratio. Indeed, although it is of no practical significance, in the limiting case of an infinite number of transmitters the interference area and the service area coincide.

There are other advantages in the use of multiple synchronous transmitters, particularly in grade of service. In the case of a single transmitter, if the radiated power is chosen so as to give a 20 dB margin at the limit of service range (for a 1% drop-out rate in Rayleigh fading), then nearer to the transmitter the mean signal power will be higher and the drop-out rate lower. If the mean signal power is increased by 10 dB the drop-out rate falls to nearly 0.1%. Bearing in mind the inverse power law relating mean power to range, this will occur when the range has been reduced by 44% with a single transmitter. In the case of four synchronous transmitters the reduction of range is around 19%, so a higher grade of service is maintained over more of the total service area. On average the receiver is closer to a transmitter for more of the time.

Multiple transmitters are more energy conserving than the same coverage with a single transmitter, and therefore cause less spectrum pollution. In the four-site example discussed above, a single central transmitter would need to have a service range

$(1+\sqrt{2}) = 2.4$ times that of the four grouped transmitters for the same coverage. If each of the latter radiates power P, the total power of the group is $4P$ whereas the single central transmitter would need to have a power of $(2.4)^4 P = 33P$ which is $+9.2$ dB relative to the multiple transmitter case. Again, the more transmitters the greater the reduction.

Finally, there is an advantage from using multiple transmitter sites in system reliability. Failure of any one transmitter will not result in total loss of service, but only in degradation of service quality over part of the service area. Also lower power transmitters can be made more reliable by rating them more conservatively.

It might be thought that the use of multiple transmitter sites is necessarily expensive, but this may not be so. For radio transmission sites which attempt to cover large areas the principal cost is often site acquisition and civil engineering work, including the cost of erecting the mast, which varies as roughly the cube of its height. Thus to choose many transmitting sites, each covering relatively small ranges from antennas at no great elevation (perhaps even on top of existing buildings), can be a cheaper alternative to high masts on hill tops, even allowing for the cost of the necessary links between sites. The visual environmental impact of a high mast also needs to be taken into account. Planning permission to erect such a mast may be difficult to obtain, and rightly so. Finally in the case of a high mast there is a real danger of line-of-sight reception (with inverse square law characteristics), which leads to intractable interference problems at long range, as has been shown. In general, the use of high masts is not compatible with spectrum conservation in VHF and UHF urban radio systems.

So far it has been assumed that the transmitter carriers are perfectly coherent, that is of precisely the same frequency and in fixed phase relationship. Provided that the scattering objects in the environment – hills, buildings and so on – are fixed in location, the pattern of scattering is unchanged with time. Plotting the signal power over the ground in the service area would reveal a complicated pattern of variation, with nulls in some locations, maxima in others and intermediate levels of signal power elsewhere. This pattern remains fixed in space, so that if a mobile user happened to try to receive when in a null, the low signal strength would persist indefinitely. To find a better signal it would be necessary to move the receiver location.

Sometimes, however, a very small frequency offset may be tolerated between carrier frequencies (still carrying identical modulation). Provided that the offset is small enough the effects are slight. Clearly they need to be less than the unavoidable doppler shift produced by the motion of mobiles in cars, which will thus dominate frequency shift effects. Thus the frequency offset:

$$\Delta\omega \leq \frac{v_{max}}{c} \cdot \omega_o \tag{9.20}$$

where v_{max} is the maximum vehicle velocity, typically 30 m/s for road vehicles, and ω_o is the carrier frequency. Thus typically $\Delta\omega$ must be less than $10^{-7}\,\omega_o$, or 0.1 Hz/MHz. This mode of operation is known as **quasi-synchronous** transmission. Provided that the frequency shifts are small, the effect of quasi-synchronous working is that the pattern of signal maxima and minima begins to move across the terrain (perhaps in a fairly complicated way if there are many transmitting sites). Thus the signal received at a fixed location varies continuously with time over the full range of values.

In some situations, for example use by the fire service, the location of the radio equipment is fixed by operational circumstances. If so, variation of the signal strength with time may be considered a positive advantage, since it means that at any location communication will be possible most of the time, but aside from special cases of this kind there is not much to commend in quasi-synchronous working. It might be thought that an advantage is that the frequency synchronizing links (radio or cable) between the transmitting sites can be omitted, but for optimal working the permissible frequency offsets – less than one part in a million – are so small that it is usually cheaper to design the link hardware for phase locking than provide free-running oscillators, such as rubidium standards, to this accuracy. Although once popular quasi-synchronous working is now much less common.

The multiple site approach is not restricted to transmitters. If two-way communication is required, multiple fixed receiving sites (possibly co-sited with the transmitters) show similar advantages in receiving transmissions from anywhere in the service area. In the simplest (but quite effective) systems the received signal is taken (by switching) from the receiver at which the S/N ratio is greatest but superior performance is achieved if the signals from several sites are combined. If this is done at RF it would be possible to use the multiple receiving sites to form an adaptive array antenna, but the inter-site link cost is high.

So much for multiple sites. An alternative approach to improving efficiency of spectrum utilization can be by careful use of directional antennas, particularly in conjunction with multiple transmitter systems. As so often with antenna engineering, relatively simple and low-cost antennas can give a useful improvement; however, a law of diminishing returns soon sets in and substantially greater investment on antennas produces only modest further returns.

In the four-transmitter example considered above, if the transmitters no longer have omnidirectional antennas a considerable advantage can be obtained. Suppose, in the first instance, that the antennas are replaced by simple two-element Yagi arrays, arranged so that the main lobe is pointed towards the centre of the square – an arrangement with only a very small additional cost compared with a single dipole. The main lobe power gain is typically 5 dB (×3.2), so while maintaining the required signal strength at the centre the transmitters may be moved radially outward, increasing their distance from the centre by 26%. Although radiation towards the corners of the square is now to the rear of the array, and is therefore lower than that towards the centre by the front–back ratio of the array, typically 10 dB, the re-siting of the transmitters means that adequate coverage is still obtained, indeed it is slightly improved. Nevertheless the power density being radiated in that direction is reduced by around 5 dB, and therefore the interference range is only about three-quarters of what it would have been with omnidirectional antennas. With more sophisticated (and more expensive) directional antenna structures – a 90° corner reflector might be chosen in this case – even better results are obtained. As the forward gain and front–back ratio improve the transmitter sites move progressively nearer the corners of the square.

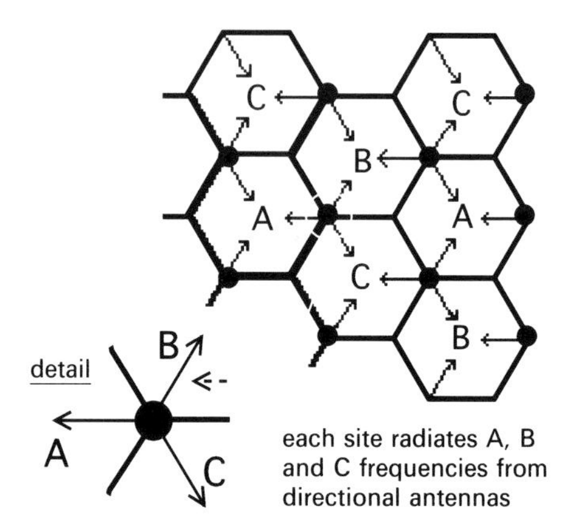

Fig. 9.7
By locating transmitters at an apex of cells (rather than in the centre) each cell is irradiated from three antennas without the need for additional sites.

Similar improvements are obtained with more complicated area coverage configurations. To achieve more economical spectrum use in conventional hexagonal tiling patterns for frequency assignment, cells may combine directional antennas with multiple transmitters. A particularly interesting case is when each cell is irradiated by three transmitters with a directional antenna (such as a 120° corner reflector) (Fig. 9.7). A configuration of this kind can be realized using no more transmitting sites than required for single transmitters at the centre of (slightly differently located) cells. Although more sophisticated antennas and three transmitters are required at each location, since site-provisioning costs are likely to be large, the additional cost of this type of installation relative to the simpler system is not prohibitive. In this case with $N = 3$ the value of q is raised to 19 dB, with comparable improvements for more complex assignment patterns.

Questions

1. In terrestrial radio systems, why does free-space propagation lead to less efficient spectrum utilization than scattering propagation?

2. A BPSK data system using omnidirectional antennas in a scattering environment is engineered for 99% probability of contact at its maximum range of 10 km. Beyond what distance can the channel be reused? (64.3 km) If the specification is relaxed to 97% probability at 10 km, what is the new reuse distance? (50.7 km)

3. Radio data transmission on a straight east–west highway is provided by a series of transmitting sites on the highway at 1 km intervals. At each site two frequencies (A and B) are radiated from directional antennas with infinite front–back ratio, one with its main lobe pointing due E, the other due W. Radio communication on lengths of highway between successive transmitting sites is provided by frequencies A and B alternately. What is the worst case mean wanted/unwanted signal ratio, assuming scattering propagation? (31 dB)

CHAPTER 10

THE SECRETS OF SEQUENCY

So far we have considered reducing interference between transmissions by ensuring that they are sufficiently far apart for the interfering signal power to be small enough compared with the wanted signal. In doing so it has been necessary to consider cases where they would be capable of interfering with each other as a matter of course, so we have assumed that they are present at the same time and are co-channel, that is in the same frequency assignment. It was necessary to make these assumptions because the timing and format of signals (so far we have only considered frequency) is another means by which they can be made effectively orthogonal (that is, non-interfering) as we saw in the development of tiling patterns. We now consider the question of format in more detail.

10.1 The frequency domain

Since radio use first began, mutual isolation between transmissions has been achieved by radiating them in different frequency bands and using conventional wave filters to separate them. The earliest receivers used single resonant circuits for this purpose, but the selectivity of this simple arrangement quickly proved inadequate and more complex filters soon began to be adopted. As the complexity of filters increased the simple **tuned radio frequency (TRF)** receiver configuration was replaced by the **superheterodyne**, which allows most of the filtering to be performed at a fixed

intermediate frequency (IF). Contemporary receivers mostly use digital filters to provide the required IF band-pass characteristic. These are cheaper by far than other kinds of filters of comparable performance and can have transfer characteristics set by software, which makes for great design flexibility.

We begin by reviewing the opportunities and problems presented by the exploitation of the frequency domain to secure orthogonality of transmissions, and subsequently we shall generalize the argument from **frequency** to **sequency**, thus taking in an additional and very important range of radio systems exploiting **code orthogonality**. As we have already seen, because radio propagation is strongly frequency dependent, it is not possible to radiate baseband signals directly and they must be modulated on to a suitable carrier so that their spectrum can be translated into the appropriate part of the radio bands. For LM this is simply a linear translation and the baseband spectrum, whether analogue or digital, is unchanged apart from its power level and location in the bands. Thus a digital signal might be modulated on to a convenient low frequency carrier using, say, QPSK and the resulting spectrum translated into a suitable radio band by RF modulation. Using LM, the RF spectrum will be of very nearly the same width as the baseband spectrum, and merely changed in centre frequency. For other forms of modulation, such as AM and FM, the spectrum is modified by the modulation process, invariably being widened, which is undesirable.

At RF the signal will need to fit its **frequency assignment**, made available by the radio regulatory authority as part of the transmitting licence conditions. This normally consists of an assigned RF channel of specified width and centre frequency (with the permitted mode of modulation also specified). However, it is commonplace for frequencies to be assigned at regular intervals, that is to say in channels of constant width, for example 9 kHz in MF broadcasting, 12.5 kHz in UHF private mobile radio and 250 kHz for VHF-FM sound broadcasting.

We have seen that for all transmissions encountering interference there exists a co-channel protection ratio, which is the ratio of wanted to interfering signal powers to give an acceptable grade of service. This is a positive decibel figure, typically from under 10 dB (data) to over 50 dB (analogue television). It is also necessary to define an **adjacent channel protection ratio**, the ratio of power in the adjacent channel to that in an wanted channel which just permits acceptable service. (Note that the definition is the *inverse* of that for the co-channel protection ratio; this is purely a matter of convention, in order that the decibel figure shall be positive.) The adjacent channel protection ratio is primarily a consequence of the technical characteristics of the receiver, particularly its **selectivity**, although to a lesser extent it is also a function of the characteristics of the interfering and wanted signals, as with co-channel protection ratio. The RF energy in an adjacent channel is much better tolerated than co-channel interference, so the adjacent channel power can be relatively large and this means that the adjacent channel protection ratio is a large positive decibel figure, for example 70 dB. How is this achieved?

First, attention must be given to the transmitter specifications. In general, there will be a specified maximum radiated power level in the transmission **passband** and the power will be required to fall to a specified level at a short frequency interval away from the edge of the transmission (Fig. 10.1). Thus between the passbands of adjacent transmissions there will be an interval, often called the **guardband**, in which any power radiated plays no useful role in signal transmission, being simply spectrum pollution. Because of

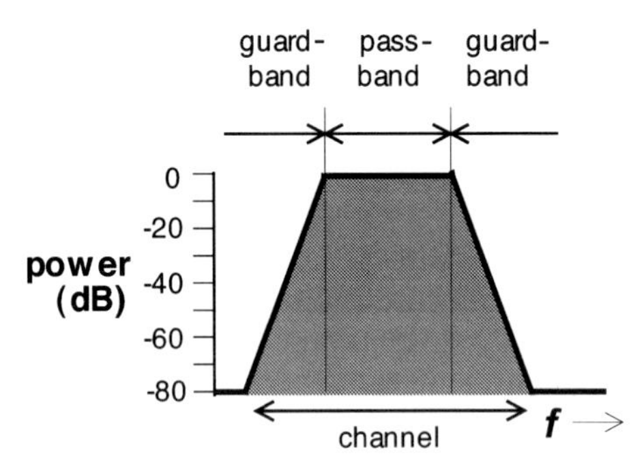

Fig. 10.1
A typical transmitter specification. Emissions must be within the shaded area.

the finite roll-off rate of filters defining the edges of the passband of a transmission and, more importantly, the generation of **spurious outputs** just outside the passband due to technical inadequacies of transmitters, the guardband has to be provided to prevent the skirts of the spectrum of spurious products from overlapping significantly into the adjacent channel. How wide the guardband must be depends on the attainable transmitted signal spectrum and the tolerable spurious signal level in the adjacent channel – ideally one would like it to be as narrow as possible to achieve best economy of spectrum use.

As for the receiver, it might be thought that a band-pass IF filter with an extremely rapid roll-off, such as is now easily attainable using digital techniques (making it possible to realize filters having in excess of 100 poles), the receiver passband at least would be very well defined. Quite aside from the problems of poor transient response in filters with such ultra-rapid roll-off, which can be eased by careful design, the situation turns out less satisfactory in practice than this suggests. In addition to providing an output from the wanted signal, receivers unfortunately also have a number of different spurious responses, which can result in a significant output from the IF stages for various interfering inputs.

We now, therefore, consider technical factors in both the receiver and transmitter which, between them, determine what the interference potential between transmissions is, and hence what guardband is required, and what adjacent channel protection ratio is needed, among other things.

10.2 Receiver limitations

Obviously it is primarily the selectivity of the receiver which isolates signals in different channels. To illustrate the vulnerability of the receiver to unwanted transmissions in the frequency domain, consider specifically a superheterodyne configuration. The incoming signal from the antenna is centred on a frequency ω_s and has a bandwidth $\Delta\omega$. (Fig. 10.2). It passes through an initial radio

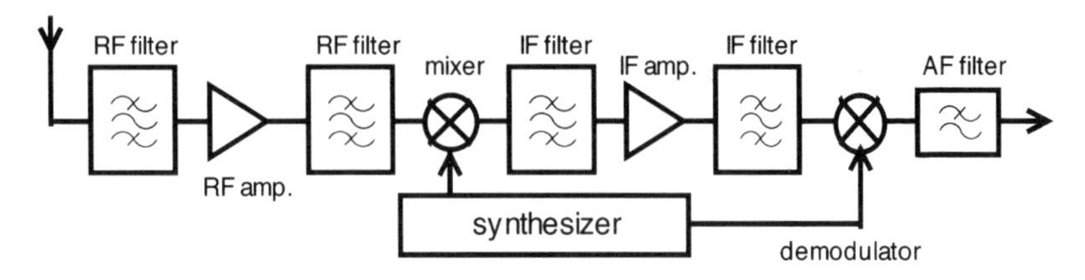

Fig. 10.2
Superheterodyne receiver configuration.

frequency band-pass filter and may be amplified before passing to the mixer. There the signal is multiplied by a local oscillator output at frequency ω_o with the result that the signal centre frequency is translated to ω_{IF} where:

$$\omega_{IF} = |\omega_s - \omega_o| \tag{10.1}$$

On the RHS we take the magnitude because the signal frequency may equally well be above or below that of the local oscillator. This at once identifies the first spurious response of the receiver. For either location of the signal frequency (above or below) there will also be an additional response at the other frequency, separated from the signal frequency by $\pm 2\omega_{IF}$. This is the **image** (or second channel) response. Since both the wanted signal and the image generate outputs from the mixer at ω_{IF} it is obvious that the only way to minimize the image response is by some kind of signal processing prior to the mixer, most often by filtering. RF band-pass filters before the mixer are designed to pass the wanted signal in their passband, while the image response falls in the stop-band and is attenuated. RF selectivity is often the only protection against image response, although image-cancelling mixers are also sometimes seen.

Since the wanted signal and its image are separated in frequency by $2\omega_{IF}$ it follows that the higher the intermediate frequency the easier it is to design RF filters to attenuate the image response heavily. However, it may be difficult to construct a suitable IF band-pass filter if the IF is too high, particularly for systems using narrow channels. Digital filters must operate at clock rates very high compared with their centre frequency. These considerations create a preference for a lower intermediate frequency. Evidently there is a design clash here; either a compromise IF is chosen, neither too high nor too low, or the double-superheterodyne receiver configuration is adopted. In a double superheterodyne there are two frequency conversions, the first to a relatively high first IF, which eases the problems of image suppression, and the second to a much lower second IF to ease filter design. In this way compromises over the choice of IF are neatly avoided, but at an obvious cost penalty.

10.3 Receiver non-linearities and retroconversion

Suitable choice of first IF frequency and good-enough RF filters can reduce image response in receivers to negligible levels. The same is not true of some other receiver spurious responses. The most important of these arise because the response of the receiver, and in particular of the **front end**, from the input (antenna) port up to the mixer, is not linear. In general it will be necessary to represent the transfer characteristic of the front end by means of a polynomial:

$$e_{\mathrm{out}} = a_1 e_{\mathrm{in}} + a_2 e_{\mathrm{in}}^2 + a_3 e_{\mathrm{in}}^3 + \cdots \tag{10.2}$$

When two signals are present:

$$e_{\mathrm{in}} = \{e_1 \cos(\omega_1 t + \phi_1) + e_2 \cos(\omega_2 t + \phi_2)\} \tag{10.3}$$

If this expression is fed into the polynomial a large number of terms are generated consisting of products of powers of cosine functions. In particular:

- the a_1 term generates no products,
- the a_2 term generates squares of each cosine and the product of the two cosines,
- the a_3 term generates the cubes of each cosine and also the product of the square of each cosine with the other cosine,

and so on. If these products of cosine functions are transformed, in the standard way, into sums of cosines with more complex frequency terms, it is easily shown that:

- the a_1 term generates components with frequencies ω_1 and ω_2 only,
- the a_2 term generates components with frequencies $2\omega_1$, $2\omega_2$, $(\omega_1 + \omega_2)$ and $(\omega_2 - \omega_1)$,
- the a_3 term generates components with frequencies $3\omega_1$, $3\omega_2$, $(2\omega_1 + \omega_2)$, $(2\omega_2 + \omega_1)$, $(2\omega_1 - \omega_2)$ and $(2\omega_2 - \omega_1)$,

and so on. Suppose that the frequency of the wanted signal is ω_w and let:

$$\omega_1 = \omega_w - \Delta\omega \tag{10.4}$$

$$\omega_2 = \omega_w - 2\Delta\omega \tag{10.5}$$

where $\Delta\omega$ is the channel spacing. Then all the second-order products are far removed in frequency from ω_w (and therefore after conversion will not generate signals within the IF passband) and so also are the $3\omega_1$, $3\omega_2$, $(2\omega_1 + \omega_2)$ and $(2\omega_2 + \omega_1)$ terms from the third-order non-linearity. However:

$$2\omega_1 - \omega_2 = 2\omega_w - 2\Delta\omega - \omega_w + 2\Delta\omega = \omega$$

So in the presence of third-order non-linearity signals in the next channel and the next but one below the wanted signal can combine together to put a spurious signal in the wanted channel. The analogous effect also occurs if the next and next-but-one channels above the wanted signal are occupied. This can be a serious problem in both cases, since the third-order coefficient a_3 must be reduced to a very small value to overcome these effects, because the signals e_1 and e_2 may be very large compared with the wanted signal. It is not difficult to show that all the odd-order non-linearities in the receiver front end can similarly give rise to spurious signals in the IF passband, but those higher than third-order terms are usually less significant.

There is nothing for it but to make the early stages of the receiver as linear as possible. The hardware details of how this is done go beyond the scope of this book, but include very careful attention to the choice of active devices for the receiver front end, for example by the use of MOS and field-effect (rather than bipolar) transistors, which can be made more linear. Because all active devices are invariably non-linear to some degree, however, it may also dictate omitting any RF amplifier before the mixer, despite the fact that this will generally worsen the noise figure of the receiver. In contemporary engineering practice, designing for heavily congested radio bands, spurious responses are frequently the limiting factor on receiver sensitivity. Noise is then a secondary consideration.

Another common cause of degradation of adjacent channel protection is **retroconversion**. Ideally, in a superheterodyne the local oscillator should have a spectrum consisting of a single line at ω_o, but in practice because an active device – a bipolar or field-effect transistor – must be used to sustain the oscillation, there will also be significant noise from this device present in the oscillator output. This will have the effect of adding 'skirts' to the spectrum, with small amounts of energy at frequencies significantly different from the centre (Fig. 10.3). In particular there will be some energy at the retroconversion frequencies:

$$\omega_{retro} = \omega_o \pm n \cdot \Delta\omega \tag{10.6}$$

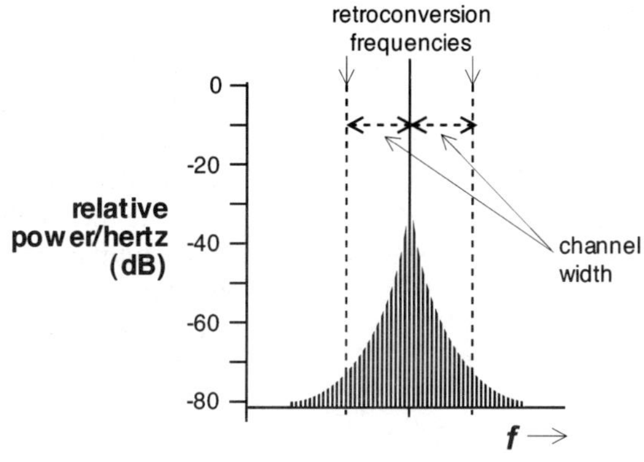

Fig. 10.3
A possible receiver local oscillator spectrum.

where *n* is an integer, although in practice values above one are rarely significant. Components at these frequencies can mix with adjacent channels to produce signals within the IF passband. What is observed is unexpectedly large amounts of in-band noise. Although the noise components of the local oscillator output are of low amplitude the adjacent channel signals may be very large, so the effect is significant.

The solution to this problem is to design the local oscillator for low noise relative to the wanted output. The 'brute force' approach to this is simply to have the oscillator work at a high power level, which increases the wanted waveform without significantly increasing the noise, so that the ratio is improved. Sometimes for this reason receiver oscillators are seen operating at power levels up to as high as 10 watts RF. It hardly needs to be said that any such approach is out of the question in portable battery powered equipment, and it can anyway cause problems elsewhere in the receiver to have so much RF power generated in the front end. Consequently active device designers have concentrated on developing components which have a superior S/N ratio even at low power levels. Generally (depending on the frequency range) field-effect and hot electron devices prove superior to bipolar transistors, and as the semiconductor on which devices are constructed gallium arsenide gives better results than silicon.

In the case of PLL (phase-locked loop) frequency synthesizers, now almost universally used as local oscillators, the waveform is generated by a **VCO** (voltage-controlled oscillator), which can be 'pulled' to the wanted frequency by a voltage derived from the PLL. This pulling voltage is a common source of spurious low-level frequency modulation, and hence of a noise-power component in the output. A low-pass filter on this line follows the phase sensitive detector, and from the noise point of view the lower its cut-off frequency the better. However, a compromise has to be made, since the response time of the PLL is extended by a lower cut-off frequency, and there is always a practical limit to how long the VCO can be allowed to 'settle' after a frequency change. Of course, this is a particular problem for receivers which have to change frequency rapidly, like frequency hopping equipments.

In summary, both non-linearity and retroconversion can result in receiver selectivity significantly worse than the passband profile of the IF filter, thus reducing the orthogonality of the receiver to unwanted adjacent channel signals.

10.4 Transmitter limitations

Transmitters have limitations much like those of receivers, but with an important difference. Whereas the deficiencies of receivers make them vulnerable to unwanted signals, transmitters actually radiate unwanted and unintended energy. In short, they cause spectrum pollution.

The radiation of transmitter harmonics can be a problem, but they should be easily and permanently suppressed by careful transmitter design to reduce harmonic content in the power amplifier output (as a result of transfer non-linearities) and if necessary the incorporation of suitable low-pass filters in the feed to the antenna. Other problems are less easily dealt with, and the worst of all also arise from intermodulation effects due to non-linearity in the power amplifier. In general the output of the transmitter will cover a range of frequencies in the assigned channel which can be thought of as a number of closely spaced spectrum lines. Odd- (and particularly third-) order non-linearities will result in spurious intermodulation products within and just outside the channel, in the same way as is discussed above in connection with receivers. Products just outside the assigned channel are particularly troublesome because if they extend beyond the guard-band they will cause adjacent channel interference.

The need to build much more linear power amplifiers is now widely recognized and a number of solutions to the problem have been found, of which **predistortion** and **feedback** are the most successful. Predistortion, as its name suggests, involves applying distortions to the modulating signal which cancel those of the power amplifier. If the latter (ignoring even-order non-linearity terms since the transfer properties of the amplifier are likely to be symmetrical) has a transfer characteristic of the form:

$$e_{out} = A_1 e + A_3 e^3 + A_5 e^5 + \cdots \tag{10.7}$$

Then provided, as is always the case, that the coefficients of the third- and higher-order terms are small, suppose that the amplifier is preceded by a network with the transfer characteristic:

$$e = e_{in} - B_3 e_{in}^3 - B_5 e_{in}^5 - \cdots \tag{10.8}$$

Clearly if:

$$B_3 = \frac{A_3}{A_1} \tag{10.9}$$

the third-order term will be cancelled and only higher-order terms will remain. Similarly a value for B_5 can be obtained which will remove the fifth-order term, and so on for all higher orders which may be significant.

At one time the predistorting network was realized in analogue form, which made it inflexible, since once non-linear elements (such as diodes) had been chosen to give suitable predistortion characteristics they could not easily be changed later. Over time the characteristics of the amplifier are likely to change due to ageing of components and active elements, so fixed predistortion has often been found only temporarily effective.

Subsequently it proved possible to use digital signal processing to carry out the predistortion. In this case the parameters of the virtual predistorting network are set by software and easily changed, which means that they can be reset as often as necessary, for example at the beginning of each operating period. Built-in test systems may be used to measure the non-linearity of the amplifier and automatically adjust the predistortion parameters to minimize these effects.

The use of **negative feedback** is an even more powerful way of reducing distortion. In the case of an RF amplifier, evidently both the angle and amplitude of the wave need to be corrected, and this

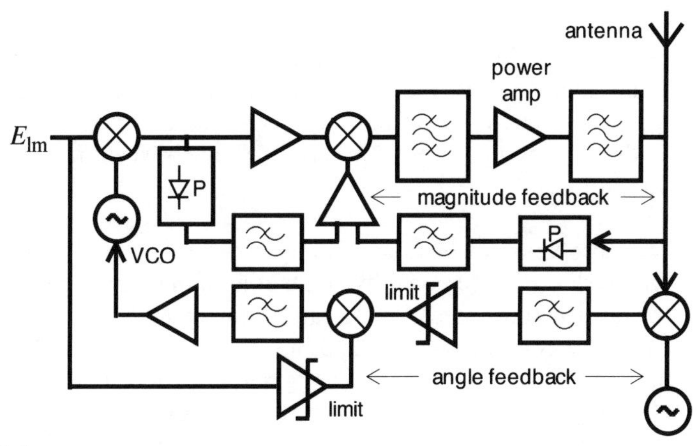

Fig. 10.4
The polar-loop feedback transmitter minimizes spectrum pollution (P = precision rectifier).

can be done by resolving the RF output into orthogonal components, either angle and amplitude (r, θ) in **polar-loop** transmitters, or the corresponding in-phase and quadrature amplitudes (x, y) in the **Cartesian-loop** form, which is the more widely used (Fig. 10.4). Negative feedback is then applied conventionally with a single-pole low-pass filter in the feedback path to ensure that the Nyquist criterion for stability is met. This filter sets the bandwidth over which the feedback is effective, usually several times the channel width. The reduction of transmitter spurious emissions can be dramatic.

Using either of these techniques (or others with comparable effect) the amplifier is linearized and its potential for spectrum pollution much reduced. However, it sometimes comes as a surprise to those designing radio systems to find that amplifiers are not the only source of non-linearity present on radio transmitting sites. It is a fact, though, that any non-ohmic junction in electrical conductors can give rise to those odd-order intermodulation effects which cause so much havoc. An example is the well-known 'rusty bolt effect'. A metal structure, such as an antenna mast, will have many metal members joined, often their surfaces being held together by bolts or rivets. Oxide or other chemical contamination of the surfaces – rust or corrosion – will usually cause the junction to be non-ohmic, and RF voltages induced in the structure will therefore generate non-sinusoidal currents in the junctions. These contain intermodulation products which are re-radiated. Co-sited receivers are likely to suffer serious interference on certain channels when the transmitters are in use.

This problem, for obvious reasons particularly troublesome in sea-going ships, is overcome by eliminating non-ohmic junctions in all metal structures used close to radio equipment. This can be done in several ways; one is by interposing 'glimp' (in the bolted-up joints, an inert silicone grease containing finely milled ceramic to give it a very high dielectric constant) so that the admittance of the joint is far larger than, and dominates, its conductance. Another obvious approach is to avoid pressure junctions by making a fully welded structure, although when this is done high quality welds are essential. Vehicle bodies made from spot-welded steel sheet are sometimes troublesome if the quality of the spot-welds is not adequate at RF. Very serious 'rusty bolt effect' can occur, even in joints that are mechanically strong, and mobile radio installations may be seriously compromised. Certain car makes are known to give this trouble, and there is no solution except to use a different vehicle.

As will be apparent from the discussion so far, with good design practice in both transmitter and receiver the guardbands required around transmissions can be minimized, leading to economical channel spacing and optimal use of the radio spectrum.

10.5 Uses of frequency domain orthogonality

Two-way radio systems are either **simplex** or **duplex**. In the former transmission is only permitted in one direction at a time and at the end of each transmission the transmitter is disabled at the transmitting end and the receiver brought into use, while the reverse happens at the other end of the radio link. Simplex voice systems are equipped with a 'press to talk' switch on the microphone (which enables the transmitter) and the person communicating must exercise the discipline of saying 'Over' (or its equivalent) at the end of each transmission, and then releasing the 'press to talk' switch, which enables the receiver. Some users welcome the discipline that this 'over-over' protocol introduces, and it is very tolerant of long time delays in transmission, useful for satellite and deep-space communication. However, most people prefer duplex voice systems, in which either party can talk and be heard at any time since this facilitates normal conversation and a 'press to talk' switch is not needed. By contrast for data transmission simplex working, with automatic 'over-over' signalling, is quite usual and duplex is rarely necessary.

For simplex systems exploiting frequency orthogonality, either one- or two-frequency working may be chosen. In one-frequency working the same channel is used for up and down transmissions, whereas in two-frequency working two groups of channels are assigned, usually with a substantial frequency interval between them. All up channels are drawn from one group and the down channels from the other.

Paradoxically, two-frequency working proves to be more spectrum efficient. This is because in the case of one-frequency simplex allowance must be made for the possibility that two mobiles very near each other are exchanging messages with the base station at the same time. But if one transmits while the other is receiving the possibility of severe interference is substantial, because they are so close to each other (inverse fourth-power law). This makes it impossible for them to be on adjacent channels, or even channels close to each other in the group. Since there can be no constraint on which mobiles go where, it becomes necessary to leave channels unassigned in order to avoid adjacent channel working, so the efficiency of spectrum utilization is poor.

Frequency domain duplex systems will almost always be two frequency, known as **frequency division duplex (FDD)**, since in this case simultaneously operating transmitters and receivers are actually co-located and it is both difficult and expensive to achieve sufficient isolation between them in the single-frequency case. As we shall see below, not all duplexing uses frequency orthogonality to separate up and down signals, however.

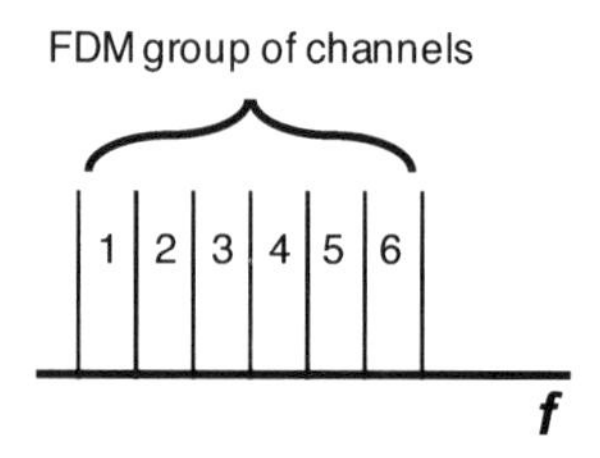

Fig. 10.5
A multiplex or group of channels is used to convey a number of distinct signals, here six. For FDMA one channel may be assigned as a hailing channel.

Sometimes it is desired to send a group of signals over a certain radio path, such as a point-to-point link, in which case frequency domain orthogonality makes possible **frequency division multiplexing (FDM)**, in which a number of channels (usually but not essentially adjacent) are used to carry one signal each of the group, in the obvious way (Fig. 10.5). Modern linear-amplifier transmitters can handle a group of channels of this kind, which can therefore be combined at low power level or generated as a group, using digital processing, and subsequently amplified to the required power level for transmission. In the receiver the group of channels is handled as a single wide-band transmission through to the IF, then individual channels are demodulated digitally. Monolithic circuits (chips) perform these functions up to increasingly high data rates.

An important variant of FDM is **frequency division multiple access (FDMA)**. In many applications, such as mobile radiotelephone, it is required that a single base station (or group of simulcasting stations) should be able to contact a number of other users, often mobiles within its coverage area. The requirement is more complicated than for broadcasting, since both **up** (mobile to base) and **down** (base to mobile) transmission paths are required and quite different signals are to be exchanged with each mobile. A widely adopted solution to this problem is to use what is in effect FDM. A group of channels is assigned for the up path and another for the down path. In the operationally simplest system one of the channels in each group, called the **hailing channel**, is kept for system management. A mobile that wishes to contact base makes a call on the up hailing channel and the base station then assigns it a vacant up channel to be used for the duration of that contact and also tells it what down channel it will be using, again choosing one which happens to be vacant at that time. The mobile then re-tunes its transmitter to the specified up channel and its receiver to the specified down channel, after which traffic can be passed in the usual way. Similarly, if the base station wishes to initiate traffic to a mobile it calls the mobile on the down hailing channel, assigning working channels as before.

Further complications of the systems are introduced to cope with problems arising from contention between base stations for the up hailing channel. Systems are also possible which do not use assigned hailing channels but exploit the working channels for system control purposes. Neither of these requirements greatly modify the basic design.

10.6 Non-sinusoidal carriers

Hitherto we have limited ourselves to considering sinusoidal carriers, but this is an unnecessary, and as it turns out undesirably limiting, constraint on the range of radio systems that can be designed. For the moment we need not concern ourselves about the shape of the carrier waveform – indeed for simplicity in what immediately follows we shall consider primarily rectangular waves (which, as already explained, can always be shaped or filtered to sinusoidal form). The crucially distinctive point about a sine-wave carrier is that the zero-crossings occur at regular intervals. The interval between successive zero-crossings is half the period of the sinusoid, so the frequency is half the reciprocal of that interval. (Specifying the waveform in terms of frequency, rather than period, is partly a matter of convention, but proves convenient because the frequency bandwidth of a signal is directly related to the rate at which it is conveying information and is thus a useful measure of the spectrum occupancy.)

However, in the more general case, the interval between the successive zero transitions need not be constant. Considering the case of white noise, for example, the interval varies continuously, and can only be described by a statistical distribution. In another case of great practical importance successive intervals between zeros are of the form $n \cdot \tau$, where τ is a fixed time interval and n is an integer which can take a specified range of values. In effect, this is a rectangular wave in which the length of each half-cycle can be varied in incremental steps. Usually after a certain number of half-cycles the sequence repeats itself, and it is thus not truly random even if there is no simple law relating each interval to the next. The possibility of using a waveform like this as a carrier for radio signals introduces new and valuable design freedoms.

How can such waves be specified? There are many possible valid answers to this question, useful for different purposes. One way of specifying such a waveform recalls that for a simple repetitive sine or square wave, we can specify the wave (without regard to amplitude) by its frequency, which is half the inverse of the interval between successive zero-crossings. Indeed, unambiguously we could simply specify inverse interval. This would be a point in a line, equivalent to a location in a **one space**. If this interval is non-constant, however, we would need a location in a space of many dimensions in order to specify it. The inverse of the first zero interval could be plotted along the first axis of such a space, the inverse of the second zero interval along the second axis, of the third along the third and so on (Fig. 10.6). (These plots could allow either continuous variables or quantized values, as appropriate.)

After a certain number of intervals the sequence begins to repeat itself, and at that point the waveform is completely specified. This therefore sets the number of dimensions that the space must have in order to specify the waveform fully by a unique location in that space. (For a truly random waveform, which never repeats, the dimensionality is infinite, so it is not usefully described in this way.) A space of this kind, actually a **hyperspace**, in which a single point defines a sequence, is known as the **sequency space**. It degenerates to the frequency domain when it becomes one dimensional, so frequency may be seen as a special case of sequency, in fact the limiting case.

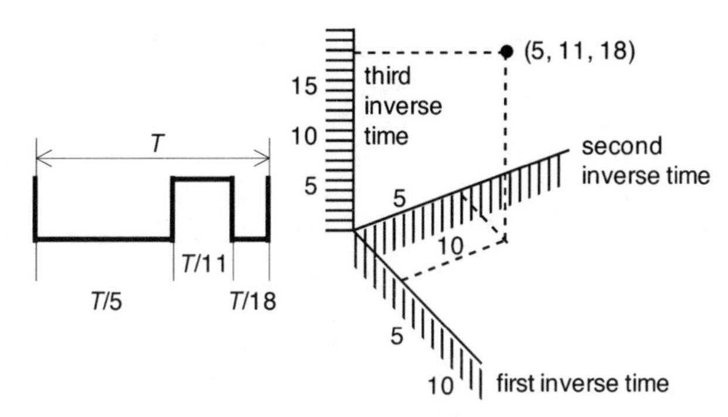

Fig. 10.6
Representing a binary sequence in three-dimensional sequency space, in. Unrealistic and for illustration only; the space is actually hyperdimensional.

As an example of how this works, a typical waveform used in radio might consist of a rectangular waveform which begins to repeat after the 101st zero. It can thus be specified uniquely by giving the inverse values of 100 successive intervals (first zero to second zero, second zero to third and so on), which would correspond to a point in a sequency space of 100 dimensions. Hyperspaces of this kind cannot be envisaged by the human mind, but even so it is possible to use them in creating mathematical models of complex situations.

Seeing that sequency is a generalization of frequency helps the understanding of the important **CDM** (**code division multiplex**) and **CDMA** (**code division multiple access**) systems because they can be seen in the light of improved and generalized versions of FDM and FDMA systems, sharing much of the same mathematical analysis and properties, but their practical usefulness much enhanced by their greater generality.

10.7 Demodulation and filtering

In the sequency domain it is possible to carry out modulation and filtering functions just as in the frequency domain. A good place to begin is with the concept of **correlation** – which is a way of measuring the similarity of two waveforms (or functions). Consider two hyperdimensional binary sequences $x(t)$ and $y(t)$ (often called **codes**) which have a common clock period τ and take the values $+1$ or -1. (We also take care that the two sequences have the same start time.) They will, of course, correspond to two points in sequency space. We pose a question: is it possible to measure how alike these two functions are?

From the product $z(t) = x \cdot y$ we can obtain a useful clue. Take an average value of the product over a time period T which is long compared with τ, then:

$$Z = \frac{1}{T}\int_0^T x \cdot y \cdot \mathrm{d}t \tag{10.10}$$

This parameter is known as the **cross-correlation coefficient** for x and y (and it could be calculated in just the same way for time-dependent functions in general and not only for the binary functions considered here).

The limiting values of the correlation coefficient are easily arrived at. Suppose that y is identical with x, that is that the two codes correspond to points in sequency space which are coincident, then whenever one of them is $+1$ so also will be the other, and the same for -1. The product is therefore always $+1$, as also is the correlation coefficient; this is evidently its maximum value in the case of two perfectly correlated functions. By contrast it is easy to see that if $y = -x$ the correlation coefficient is -1, which is the lower limit, corresponding to two codes which are the exact opposite, of each other. (In this case the two codes have the same sequency but one of them is multiplied in amplitude by -1.)

If the two functions have no relationship to each other at all, that is they are of completely different sequency, whether x is positive or negative there is a 50–50 chance that y will be the same, and an equal chance that it will be the opposite. Thus the product will be either $+1$ or -1 with equal probability, and over a long period this will average to zero, the correlation coefficient for two completely

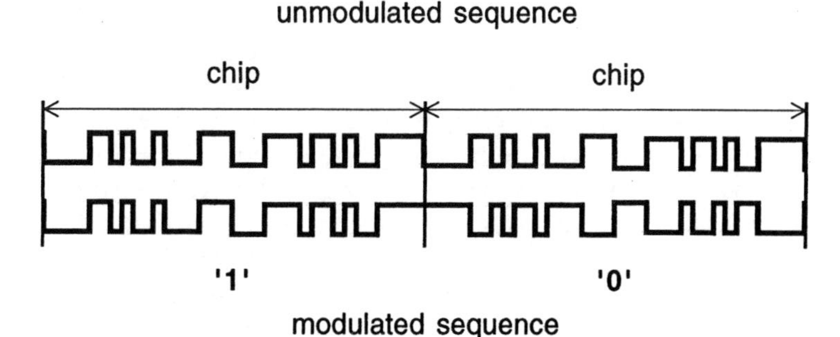

Fig. 10.7
Digital transmission using one chip per signal bit. Chips are inverted for 0, uninverted for 1.

unrelated, incoherent functions. Codes that have this property are **orthogonal**.

This leads directly to the notion of **sequency domain signal processing**. First we look at modulation. As we have seen with analogue signals, the simplest kind of modulation is that in which the message and carrier are just multiplied. However, if we assume that both are binary digital sequences it is not necessary to have a true multiplier since the modulated signal can be produced by a simple logic operation. In that case, supposing the modulating sequence positive for a logic **1** and negative for a logic **0**, modulation is equivalent to transmitting the carrier sequence in its normal form when the modulating waveform corresponds to a **1**, but transmitting it inverted when the modulation is a **0**. (Fig. 10.7).

Thus we could treat x, above, as a carrier for a baseband signal (like the familiar sinusoidal carriers in previous sections) switching it between positive and negative values for **1** and **0** modulating digits, so that if $e(t)$ is the binary modulating signal, its clock period many times longer than τ, the modulated sequence is $e \cdot x$. This type of modulation is the sequency domain counterpart of BPSK.

For a demodulator, all we need is a multiplier and a low-pass filter, such as an integrator. (The integrator is easy to model mathematically, of course, but it is not the optimum filter, as we shall see.) The incoming signal x is multiplied by a locally generated sequence y and the product integrated to give an output Z. If x and y are of the same sequency, Z will be non-zero but if the two sequencies are very different, are orthogonal indeed, the output will be zero. Thus we have a sequency-domain demodulator. What if the sequencies of x and y are only slightly different? In this case the two sequencies will correlate positively or negatively over considerable stretches of time, and the form of Z will be a new code, the difference sequence. After integration the output will depend on the integration time. If this is very long the output will smooth to zero, but if it is much shorter it will correspond to a partially integrated binary sequence.

How can sequency-domain demodulation be used for communication? The output Z of the demodulator will follow the baseband signal $e(t)$ but will not be a rectangular waveform due to the action of the integrator, tending instead to a triangular wave. Substituting a more sophisticated low-pass filter for the integrator can allow a rise time compatible with a reasonable representation of $e(t)$ whilst at the same time suppressing shorter transient outputs from the multiplier at the difference sequency.

So far we have considered only a sequency domain modulator and demodulator. To create complex signal processing other system elements are necessary, and, as it proves, entirely possible. To create a **band-pass sequency filter**, for example, we must remodulate the output Z with the local sequence y, and to achieve any useful filter response replace the integrator by a realistic low-pass filter. Such a filter will pass codes only of sequencies contained within a small volume around the point in sequency space corresponding to the local code y. The analogy with a band-pass filter in the frequency domain is exact: the frequency filter will only pass signals of frequencies within a small interval on either side of the centre frequency. By going to the sequency domain we have simply generalized the filter from a one-dimensional to a hyperdimensional form. These are just a few examples; in fact all the methods of signal processing familiar in the frequency domain have their counterparts in the sequency domain.

10.8 Orthogonal codes

Pairs of codes have been defined as orthogonal if their cross-correlation coefficient is zero. They define two different points in sequency space.

In the limiting one-dimensional case different frequencies, defining different points along the frequency axis, are necessarily orthogonal. Since the earliest days of radio, the receivers have used selective demodulators or bandpass filters to isolate wanted from unwanted transmissions, where these are made mutually orthogonal by being centred on different carrier frequencies. However, there is another way in which carriers in the frequency domain can be rendered orthogonal, even though they are of the same frequency. If one carrier is displaced in time from the other by half of the time interval between successive zeros (corresponding to a quarter of the total wave period, or a 90° angle shift in the sinusoidal case) the carriers will be orthogonal by reason of the time shift. This is exploited in QPSK modulation, as we have already seen. If signals are to be separated, though, their carriers must be orthogonal, one way or another.

Similarly in the sequency domain, low-pass demodulators or band-pass sequency filters can select out wanted signals, with the advantage that the number of axes available for selectivity is far greater than in the special case of frequency. However, here too the various carriers must be orthogonal if signals are to be distinguished, and this can be achieved either by their being different in sequency or being suitably time delayed (Fig. 10.8).

As we have seen, the sequency of a binary wave is defined as the point in space which specifies all the intervals between successive intervals up to the point at which the waveform begins to repeat. The time to repeat, which is the sum of all these intervals between zeros, is known as the **frame length**. By definition, within the frame length the sequence does not repeat. Thus a binary waveform of identical sequency but shifted in time by less than a frame length must be close to orthogonal. So:

$$\int x(t) \cdot x(t + n\tau) \cdot \mathrm{d}t = 0 \qquad (10.11)$$

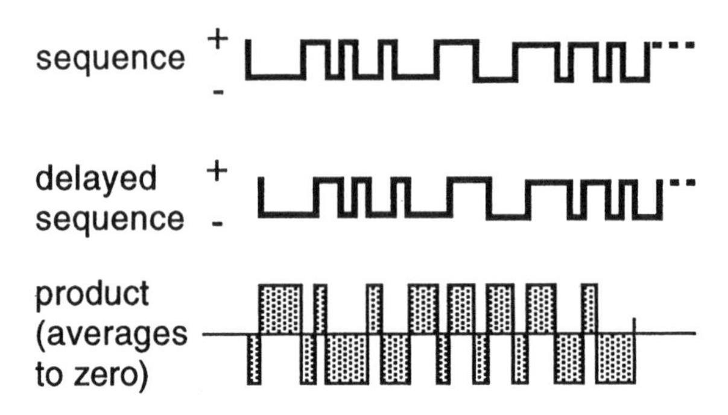

Fig. 10.8
A pseudo-random sequence and a delayed version are approximately orthogonal.

or nearly so, provided that the time shift is less than the frame length. This is one convenient way of generating approximately orthogonal sequences, by time delaying one sequence relative to the other by an interval less than the frame length. Alternatively, it is possible to use sequences (codes) with the same start time but different patterns of **1**s and **0**s so that they are close to orthogonal.

In the early days of sequency domain communication it seemed important that orthogonal codes should be capable of being generated by relatively simple mathematical algorithms, so that they could be computed within the equipment. With improvement in ROM availability this is no longer the case and codes are invariably downloaded to the equipment and held in memory. Various families of near-orthogonal codes have been worked out for this purpose, the most widely known of which are the **Gold codes**.

Thus, in outline, a sequency domain communication system can be designed using modulation on to a suitable code as carrier, transmission in that form and demodulation in the receiver. Just as with sinusoidal carriers we can use frequency orthogonality to design FDM and FDMA systems; similarly in the sequency domain we can design exactly comparable **code division multiplex (CDM)** and **code division multiple access (CDMA)** systems, both of which have proved to have marked advantages in many applications. Because the number of distinguishable sequencies is far greater than the number of frequencies, in principle a very much larger number of systems can work within interference range of each other than if we restrict ourselves to the frequency domain. This is a very substantial advantage, yet in achieving it there are inevitably some complications and problems, and it is to these that we now turn.

10.9 Complexities of sequency domain communication

Some early proponents of sequency domain communication gave a somewhat distorted view of the topic. Whilst it is true that sequency is a valid way of describing a non-sinusoidal carrier, and that the various necessary signal processing functions can be carried out in the sequency domain, this does not mean that systems can be engineered in the sequency domain without concern for their interaction with existing services, nor are the laws of radio propagation (which are highly frequency dependent) somehow circumvented by sequency domain systems. As a consequence, although location in sequency space is sufficient to define a code, it does not in itself fully describe all its important properties.

If we consider a particular binary sequence, as well as defining it by a point in sequency space, it can also be described by the Fourier series of the sequence. For example, a simple square wave $y(t)$ which is of period T has the spectrum:

$$y(t) = \cos\left(\left[\frac{2\pi}{T}\right]t\right) + \frac{1}{3}\cos\left(3\cdot\left[\frac{2\pi}{T}\right]t\right) + \frac{1}{5}\cos\left(5\cdot\left[\frac{2\pi}{T}\right]t\right) + \cdots \qquad (10.12)$$

A less regular waveform has many additional spectrum lines grouped around these, and binary modulation will spread the spectrum still further, as we shall see.

The first point to make is that since the waveform is defined by its zero-crossings, the terms above the first are not essential, since they merely determine the wave shape and do not change the zero-crossing times. If radiated they do, of course, constitute spectrum pollution, so it is desirable that they be removed by wave shaping or low-pass filtering, and normally both, leaving the first term only, a sinusoid. The same is also true of a more complex code, which is converted to a near-sinusoidal wave of variable period and will therefore have a spectrum covering a band of width dependent on the modulation clock rate. This is exactly as described above for modems in Section 7.1 above.

All of which leads to the conclusion that for any sequency domain system there will exist a well-defined band of frequencies within which energy is radiated. If the whole radio spectrum were unoccupied and uniform in properties this would be a matter of indifference, and the design of the system could be pursued entirely in the sequency domain, as some of the early authors represented. However, this is by no means the case.

First there are very many existing services which could suffer severe interference, yet are not protected at all by sequency domain orthogonality. For them the sequency domain systems are simply interfering energy, and they must be protected in the conventional way by using spatial separation or a different frequency band. Secondly, the properties of the radio spectrum are far from uniform, indeed radio energy propagates very differently in different frequency bands, as explained in Chapters 2 and 3. The ELF to MF bands are characterized by surface wave propagation, MF and HF by ionospheric sky waves, VHF and UHF typically propagate either as free-space paths or via the mechanism of scattering near the Earth's surface, SHF is a free-space band and as we move into the EHF region phenomena of atmospheric absorption are increasingly significant. Thus the frequency band used for a radio system has to be selected to secure the type of propagation most suitable to the particular service required. This implies confining the spectrum of any sequency-domain transmission to the radio band for which it is best suited.

In consequence when designing sequency-domain (particularly CDM or CDMA) systems it is essential also to keep track of the radiated signal spectrum in the frequency domain, to ensure that it is located in the band needed for the appropriate propagation characteristics, and also that it does not cause interference to other users. This is done by choosing a code as a carrier which has its centre in the appropriate part of the radio frequency spectrum and is of such a bandwidth (when modulated) as not to cause unacceptable interference with other users. The need to keep both the frequency and the sequency consequences of a design in mind together has sometimes led to difficulty for those trying to understand systems of this kind, but it is crucially important. We now examine this problem in typical situations of use.

10.10 Sequency-domain systems in the real world

In real applications the sequency domain transmission is restrained to a band of fixed width, centred in that part of the spectrum required for suitable propagation characteristics. It will frequently be more convenient to generate the spectrum in some lower band, however, and then to translate it to the desired centre frequency, using LM.

The bandwidth of the radiated signal is that of the code used as carrier (extended somewhat by the modulation) and can therefore be chosen arbitrarily. It is always greater than that of a comparable frequency-domain system, the one-dimensional case which sets the lower limit on bandwidth. In general the spectrum occupancy increases with the dimensionality of the carrier sequence, or to put it another way, with its frame length. Because the bandwidth is always greater than that of a frequency-domain system, sequency-domain systems are often called **spread spectrum** systems, and usually **direct sequence** spread spectrum systems, because the spreading is a consequence of the use of a non-sinusoidal sequence as the carrier.

Co-channel interference in spread spectrum systems has some characteristic features. Consider first the effect of received radio energy which is not correlated with the carrier sequency. The correlation coefficient is the long-term average of the product of the interfering signal and the carrier sequency; when we say that they are uncorrelated it means that this long-term average is zero. Even so, the instantaneous value of the product may be large and of either sign. For example, if we consider two completely uncorrelated binary sequences, perhaps even with different clock intervals, there will be times when both are of the same sign, giving a positive product, and others when they will be of opposite sign, resulting in a negative product. These events will occur irregularly, as dictated by the two sequences, and the product will therefore be a random binary function. The same is true if the interference is a sinusoid, whether or not modulated. Thus the ideal sequency demodulator converts uncorrelated co-channel interference power into noise power following the multiplication process. If the initial bandwidth is B hertz, after the demodulation process the correlated signal is converted coherently to a smaller bandwidth extending in frequency from 0 to B_s while the interference is converted into noise extending from 0 to $B/2$, in this case the addition being incoherent. The low-pass filter following the multiplier passes all the wanted signal power but cuts off the noise power above B_s which results in an improvement in signal-to-interference power ratio at the output of the demodulator compared with that at the input:

$$G_p = \frac{\left[\frac{P_s}{P_i}\right]_{\text{output}}}{\left[\frac{P_s}{P_i}\right]_{\text{input}}} = \frac{B}{B_s} \tag{10.13}$$

where P_s and P_i are the signal and interference powers respectively.

This improvement in signal-to-interference ratio, equal to the ratio of the spread spectrum bandwidth to the actual baseband signal bandwidth, is called the **processing gain**, and is a very useful property of sequency-domain signals. Of course, the same gain applies to the improvement in S/N ratio if no interference is present, since noise is precisely equivalent to random interference.

To show just how powerful spread-spectrum processing gain can be, we recall that QPSK data transmissions can achieve a transmission rate of 2 bits per hertz of bandwidth and requires 11.4 dB S/N ratio for 10^{-4} error rate. At a data rate of 100 kb/s the required bandwidth is therefore 50 kHz. However, if this signal is spread over 100 MHz before transmission, which would be entirely practicable when using the SHF or EHF bands, the processing gain is 2000 or 33 dB. Thus such a system could achieve the 10^{-4} error rate with a *negative* S/N ratio, actually (11.4–33) = −21.6 dB. Such a signal would, of course, be undetectable unless the correct code for demodulation were known.

Lower transmitter powers are possible with sequency-domain working, with the obvious benefits in battery- or solar-powered equipment. This apparent advantage requires careful further consideration, however. Processing gain makes it possible to use a lower power transmitter than would otherwise be needed to establish an acceptable S/N ratio in the same transmission bandwidth, but of course if the sequency-domain signal were replaced by a narrowband frequency domain signal, which similarly restricts the noise power by frequency-domain filtering, it would also be possible in this case to reduce the transmitter power comparably. In that case, however, the privacy and other advantages of sequency-domain transmissions would be lost. Nevertheless this is sufficient to demonstrate the close similarity of frequency- and sequency-domain systems, the former being no more than a special, restricted case of the latter.

Alternatively it is often claimed that for spread spectrum signals larger path losses can be tolerated in radio links, but again this is only true if similar bandwidths are compared. If the narrowband frequency-domain alternative were adopted similar advantages would be obtained. However, narrowband systems require great stability in frequency standards and filter characteristics if the signal is not to drift outside the passband of the system, a problem largely overcome by the use of spread spectrum.

This tolerance for large path loss is particularly useful in millimetre wave systems, where atmospheric absorption is a serious problem but very large bandwidths are available. Even in the 60 GHz oxygen absorption band (the worst case) the processing gain of 33 dB quoted in the example above gives an extra 2.3 km of range (typically almost a 50% extension) for marginal additional cost, while in the EHF ‘windows’ the advantage is very much greater.

Problems of frequency stability preclude narrowband solutions in this radio band.

An alternative way of exploiting processing gain is to overlay signals within the same RF bandwidth. If N different signals having mutually orthogonal carrier sequences are present in the same RF bandwidth, that is are overlaid one on another, and we can assume that the signals are large enough that the effect of noise can be ignored, then after demodulation the signal-to-interference ratio is:

$$R = \left[\frac{P_s}{P_i}\right]_{output} = \frac{1}{N-1} \cdot \frac{B}{B_s} \tag{10.14}$$

hence:

$$N = \frac{B + RB_s}{RB_s} \tag{10.15}$$

Taking the example quoted above (of a 50 kHz signal spread to a 100 MHz RF bandwidth and requiring a signal-to-interference ratio after demodulation of 11.4 dB) the nearest integral value of N is 145. This means that if 145 such signals were overlaid on each other, any one of them could be demodulated from the resulting RF spectrum with the required signal-to-interference ratio.

This calculation can, however, be somewhat misleading, since it assumes that all the signals are of equal strength. In real world situations, nearer transmitters will produce a much larger power at the receiver input than more distant ones, particularly in an inverse fourth-power propagation environment, sharply reducing the number of users whose transmissions can be overlaid. This is the **near–far problem**, which presents serious difficulties for spread-spectrum systems. It is dealt with by **power control** of transmitters, limiting the emitted power to the minimum which will just suffice for the desired communication path. Effective power control is essential to successful designs using sequency-domain carriers.

When first described the possibility of overlaying signals in spread-spectrum systems was regarded as having great potential for more effective spectrum utilization. However, we note that if each 50 kHz spectrum were assembled side by side in the required RF band, having been translated upward in frequency by LM, the total spectrum occupancy would be 31.5 MHz plus guardbands of (say) 40 MHz with a better signal-to-interference ratio, so this is a useful but not particularly spectrum conserving technique.

Sequency-domain systems show particular advantages when the signal spectrum can be spread by a large factor, giving a big processing gain. If the spectrum spreading is less, the distinctive characteristics are progressively lost, until the system approximates toward a frequency-domain version as the limiting case is approached. In considering sequency-domain systems it is also quite often necessary to partition the processing gain between more than one function, for example to permit a modest degree of overlaying and at the same time a useful improvement in the post-demodulator signal-to-interference ratio for a given low transmitter power.

10.11 The sequency domain: summing up

Some of the advantages claimed for codes as carriers (against sinusoids) are a misunderstanding, because often there are FDM solutions with properties quite similar to CDM solutions. For example, the processing gain has its exact counterpart in the frequency domain. If a narrowband signal is contained within a broad band containing noise or interference, an exactly comparable advantage will be obtained by narrowband filtering the wanted signal. Similarly, as we have seen, there is no advantage in spectrum utilization to be obtained with overlaying CDM signals compared with putting narrowband LM signals side by side. Properly made comparisons also show no advantage in transmitter power levels or tolerable radio path loss. Indeed, it should come as no surprise that what can be done in one domain generally has a counterpart in the other, because, after all, sequency is simply a generalization of frequency from one dimension to many.

In practice, though, narrowband frequency domain solutions may require more difficult hardware, such as more stable clocks as references for frequency synthesizers, and stable filters too. Moreover from HF upward, in all those radio environments where multipath propagation, and hence selective fading, can be significant, CDM spread-spectrum systems have a notable advantage over FDM, provided that the spectrum is wide compared with the coherence bandwidth (which in practice at UHF will be 2–5 MHz indoors, up to 1 MHz in the urban outdoors and up to 10 MHz for satellite links) (Sections 3.7 and 6.6). For sequency-domain signals although some components of the signal may be lost in fading nulls, most of the transmitted waveform will be received successfully. In particular the sequency will be preserved, even if minor errors in zero-crossings are caused. There is no conceivable pattern of selective fading which can cause a shift to an adjacent sequency; all that happens is a reduction in S/N ratio. Thus suitably designed sequency-domain transmissions can be more resistant to selective fading – one of the most common of all radio channel impairments – than any others.

Important though it is, this is by no means the only advantage of sequency-domain systems. For one thing, the benefits of spread spectrum for covert use were immediately perceived by the military, although civil users have also come to value the greater privacy that results. Another major, possibly overwhelming, advantage of codes over sinusoidal carriers is that there are so many more of them available. With comparable sophistication of hardware, many more locations can be distinguished from each other in sequency hyperspace than can be separated in the one space of frequency. What is more, whereas in all countries operating frequencies are allotted or assigned by governments (through a radio regulatory authority), codes can usually be chosen by system designers without reference to outside agencies. Sequencies are assigned in tiling patterns, just like frequencies, but it is no longer necessary to go to great lengths to minimize the number used, so large protection ratios are easily available.

A further important advantage of CDM and CDMA systems is their graceful degradation. Whereas with a side-by-side narrowband system there is a fixed number of channels in which signals can be accepted and no more, the penalty for overlaying one too many spread-spectrum transmissions is only a reduction in S/N (or signal-to-interference) ratio for all – invariably greatly preferable to system collapse.

For military systems spread spectrum is particularly attractive. The noise-like spectrum of the transmissions makes interception difficult, and the hyperdimensional character of the sequency domain brings real advantages against electronic countermeasures. A jammer is forced to spread energy over a wide frequency band and yet only a small part of it appears as noise in any one sequency band, corresponding to an individual transmission. The effect of jamming is therefore a modest increase in error rate for all users (which may be tolerable) whereas in narrowband systems all the jamming energy can be concentrated against a single critical channel, with possibly devastating effects.

These factors taken together are sufficient to transform the radio system design process for civil and military users alike, and together with improvements in hardware implementation (such as the ease with which LM can be used to translate the signal to the required RF band) explain the rapid growth in use of CDM in the satellite service and CDMA for cellular radio telephone, as well as a keen interest in spread spectrum for military communications.

10.12 Frequency hopping

Spread-spectrum systems are not exclusively of the direct sequence type. An alternative approach is **frequency hopping**, pioneered by the military but now with much wider application. A frequency hopping system is a fairly narrowband transmission at any one time, ideally produced by an LM transmitter, but the centre frequency of the transmission changes instantaneously at intervals, taking the transmission to different channels in turn. The frequency hop may occur only a few times per second or at much faster rates, up to at least thousands of times per second. Thus the energy transmitted is spread over a range of frequencies, although in this case they do not necessarily have to form a continuous band.

The succession of frequencies in the hopping process is usually set by a **hop sequence** stored in the equipment, usually loaded into

RAM, and is made to appear as nearly as possible random to the casual observer. Like the sequences used as non-sinusoidal carriers in direct sequence spread spectrum, the hop sequence can be represented as a point in a hyperspace, **hopping space**, although in this case the axes of the space consist of the first frequency in the sequence, the second frequency, the third and so on. Separable points in this space correspond to different sequences.

Obviously the transmitter and receiver must both follow the same hop sequence in order that they may communicate, and they must be at the same point in the sequence, that is they must be **synchronized**. Early systems sent a timing signal (possibly encrypted) by means of an additional non-hopping 'hailing channel' but this is no longer necessary. Synchronization has become easier now that accurate clocks – to a few parts per million – are cheaply available (and to be found in every quartz wristwatch) while, even more strikingly, absolute time can be obtained from the GPS-Navstar satellite system to a fraction of a microsecond, worldwide.

In some cases it is enough to have knowledge of the absolute time at which sequences start, and thus to begin the receiver sequence in synchrony from the start; however, this is only really practical for slow hopping. Otherwise it is common practice to start with a small time error and then allow the receiver sequence to 'slip' in time (for example, by temporarily running the receiver sequence marginally faster through missing out its last state) and looking at the received signal amplitudes, which become uniformly high when synchrony is achieved. At this point the sequence is returned to its correct length and the signal and receiver sequences remain in step.

The advantages, and problems, of frequency hopping systems are fairly evident. It originally commended itself to the military because it is very difficult to jam. Using radio energy to jam any one channel will be largely ineffective, since data will be lost only when the system has hopped to this one channel and at other times reception will not be impaired. There are only two effective jamming strategies: **barrage jammers** and **follower jammers**. A barrage jammer emits jamming radio energy over the whole range of channels over which the hopping sequence takes place, thus aiming to block all channels. However, for this to happen a certain minimum jamming power must be received in each channel. If the jammer is distant from the target receiver very large transmitted jamming powers will be needed to implement this strategy, with the risk of impairing own-side communications in the same band, since own-side receivers are likely to be nearer than target receivers.

The difficulty of effectively inhibiting target receivers without endangering own-side communications is further exacerbated by the inverse fourth-power propagation law which will hold for mean signal levels in the scattering electromagnetic environment characteristic of the land battle. This is the major problem faced by barrage jammers, and even the most directional antennas feasible are an inadequate solution to the problem. Many alternatives have been tried. The radio-conflict doctrine of the former Soviet Union was to mount barrage jammers in helicopters which were then to fly over opposing receiver positions, but in battlefield conditions high rates of aircraft loss were expected from this tactic. Expendable battery-powered barrage jammers placed or dropped near the target receiver positions seem a more practicable alternative approach. Much nearer the enemy receivers than 'friendly' ones, expendable jammers have the potential totally to inhibit radio use in the land battle.

Follower jammers, by contrast, are supposed to detect the hopping transmissions very rapidly and hop the jammer frequency to the same channel. Evidently this is much more economical of transmitted power, but very difficult to do at high hopping rates, and the technique has only rarely been demonstrated in practice. An additional problem arises if there are other transmissions in the band as well as the hopping transmissions to be jammed, in which case after each hop it is difficult for the follower jammer to identify which transmission is its target. This opens up the possibility of using decoy transmissions to confuse follower jammers. In very congested conditions, such as in the HF band, with many 'innocent' transmissions present, quite slow hopping (such as 5 hops/s) is remarkably secure against both follower jammers and interception. This very effective tactic is sometimes called 'hopping in the thicket'.

Although frequency hopping might seem very wasteful of spectrum, since all of the channels over which the system hops should ideally be free from transmissions, where there are a number of users their hop sequencies can be co-ordinated. Although the same set of channels is used by all users, no two hop to the same channel at the same time. In theory this can be as economical in spectrum use as a non-hopping system, with the number of channels equalling the number of users, but in practice such **co-ordinated hopping** systems (known as 'Greek dancing') are rarely seen, although they do make follower jammers almost entirely useless.

In civil applications, the security against jamming conferred by frequency hopping translates into resistance to accidental interference, and this is seen as a major advantage of the system. Long pseudo-random hopping sequences also confer a degree of privacy, although this is far short of absolute security, since recorded hopping transmissions can be reassembled after the event, given time. Nevertheless although inadequate for the military, the privacy of hopping transmissions is very acceptable to the civil user.

Frequency hopping can also give the desirable insensitivity to selective fading, caused by multipath propagation. Even if the transmission is so unfortunate as to hop to a frequency where there is a transmission null, the signal is lost only during that hop, and reappears on the next. The break in transmission, if short, can be dealt with by the usual forward error-correcting codes or repeat-back strategies (ARQ).

Discontinuous transmission is an inevitable feature of frequency hopping. It is not permissible to hop the transmitter frequency while it is at full power, since this near-instantaneous frequency modulation would generate powerful sidebands that could interfere with other users. A few early HF frequency hoppers neglected this point and consequently produced severe regular interference pulses, which became known to operators as 'the wood-pecker'. It had the effect of very clearly revealing their presence – disastrous from a military standpoint. Modern frequency hoppers reduce the transmitter output power to zero, then hop, then increase the power back to working level, controlling the rate of ramp-up and ramp-down in order to minimize transient radiation. LM transmitter designs, which intrinsically give linear amplitude control, are particularly suitable for this service.

In the case of digital communication the discontinuous transmission is no very great problem, since the data can be sent in blocks of convenient length. Analogue transmission, particularly speech, presents more of a difficulty. Since ramp-down, hop and ramp-up can easily be accommodated in two milliseconds, slow systems hopping a few times per second make no provision for the audible effects on speech, a background 'ticking' sound being tolerated by the operators. Little intelligibility is lost. Intelligibility falls as the hop rate increases (approaching the syllabic rate of speech), reaching a minimum around 15 to 20 hops per second. The only solution is to break the speech into segments, time compress them, transmit in the hopping mode, and then re-expand and reassemble the speech segments in the receiver. It is probably easier, overall, to digitize the speech in the first place and transmit blocks of data.

Crucially, however, there is an effect on the rate of transfer of data which can set an upper limit on hop rates. Suppose that the transmitter powers down, taking a time τ_{down}, then powers up again immediately afterward, in time τ_{up}. Transmission of signals is only possible for part of the time, and if the system hops h times per second, it has a transmission efficiency (ratio of information transmitted to that if transmission were continuous) of:

$$\varepsilon = 1 - h(\tau_{\text{down}} + \tau_{\text{up}}) \tag{10.16}$$

This is the principal disadvantage of hopping spread-spectrum systems, from which direct sequence systems, transmitting continuously, do not suffer. In early designs the down and up times exceeded a millisecond for narrowband transmitters of moderate power, so for a 50% reduction in transmission capacity the hop rate could not exceed 250 per second. Although these up and down times can now be greatly improved, power amplifier design begins to get difficult for hop rates much exceeding a few thousand per second, which requires these times to be reduced by two orders of magnitude.

A particularly effective frequency hopping technique for digital transmission is 'one bit per hop'. Two simultaneous hopping sequences are stored at the transmitter, one for **0** digits and the other for **1** digits, and the transmission is made at the appropriate frequency according to the digit to be sent. The receiver monitors both of the instantaneous channels and records the received digit accordingly. This is a powerful technique, very hard to intercept and robust against interference or jamming, since energy present at both frequencies gives warning of error. Without any S/N ratio penalty it can be extended to give two bits per hop (or further) by using four (or more) hop sequences, but the technique has not evoked much interest so far as is known.

10.13 Spread-spectrum systems compared

Both frequency hopping and direct modulation on sequency domain carriers (codes) produce spread-spectrum signals, but the two systems have properties which are different in important ways.

In the case of direct sequence, any radio energy within the system pass-band will produce an output from the demodulator, although if it does not correlate with the code carrier it will appear as noise. Thus other, non-coherent, CDM signals will produce demodulated noise, as will conventional LM, AM and FM transmissions and even unmodulated carriers. Thinking of the spread-spectrum pass-band as an ensemble of narrowband channels side by side, it makes no difference in which of these channels the interfering signal appears, or how many of them; in all cases the interfering power is simply converted into post-demodulator noise.

By contrast, a narrowband hopping signal can be envisaged as hopping between these channels in turn. Transmission errors only occur when the interference power in the channel to which the system has hopped at a particular time is sufficiently large that the necessary co-channel protection ratio cannot be achieved. In this case the response to narrow- and broadband interfering signals is quite different. A broadband interfering signal will affect the wanted signal on all hops (as it would with a direct sequence system) but since the available power is spread over all channels and the receiver is sensitive only to one at any one time, there is a processing gain exactly as in the direct sequence case, given by:

$$G_p = \frac{\left[\dfrac{P_s}{P_i}\right]_{\text{output}}}{\left[\dfrac{P_s}{P_i}\right]_{\text{input}}} = \frac{B}{B_h} \tag{10.17}$$

where the symbols are as before, except that B_h is the bandwidth of the hopping channel. Errors only occur if the signal-to-interference (or noise) ratio is worse than the protection ratio minus the processing gain.

For narrowband interference, equal to or less than a channel in width (including unmodulated sinusoids), interference occurs only when the system hops to the channel where the interfering power is present. When that happens, as already noted, there is no processing gain and transmission is corrupted (though only for the duration of that hop) when the signal-to-interference ratio falls below the necessary co-channel protection ratio. However, all other hops in the sequence will be unaffected by this narrowband interference. Obviously, if the presence of the interfering signal is known in advance it is possible to omit the occupied channel from the hop sequence, and in that case no interference at all occurs. This is a considerable advantage for frequency hopping, but only when the interference is predominantly narrowband. Some 'intelligent' frequency hopping systems can even adaptively modify their hop sequence while operating, to defend themselves against a changing narrowband interference environment.

Frequency hopping systems need not be limited to continuous blocks of spectrum, because the hop sequence can specify only channels known to be available, omitting others where the spectrum is already assigned to other users. Thus it can be used in relatively crowded bands where extended blocks of spectrum cannot easily be freed up; the limiting case is 'hopping in the thicket' where the system uses whatever free spectrum is available among many existing transmissions. The ability to assign spectrum with this degree of flexibility is a considerable advantage of hopping systems.

Finally, hopping and direct sequence systems behave quite differently when there are multiple users in the same band. In direct sequence systems, each new user in the band increases the post-demodulator noise a little for all other users. Thus users can be added until the attainable S/N ratio is no longer acceptable, but there is no sharp 'cliff edge' at which the system grossly malfunctions, and degradation is therefore graceful. With a hopping system, different users use different hopping sequences that at no time put two users into the same channel, so there is no degradation at all for any user up to the point at which all available channels are permanently occupied – 'Greek dancing'. There is then no further hop sequence available which will not hit one of the present users on each hop, causing serious interference and which anyway will not yield a further usable channel since it will itself suffer interference at each hop. For a hopping system, there is thus no degradation with increasing number of users up to a sharp and fixed limit where the number equals the number of available channels, at which point no further increase is practicable – the system goes over a cliff edge, like a non-hopping FDM system.

It will be seen that direct sequence and hopping spread-spectrum systems have different technical characteristics, and are thus not directly competitive but best suited to different use environments. Neither is as efficient in spectrum use as optimally designed LM systems (which come close to the theoretical Shannon limit) but their other properties may make them a better choice in a particular use environment, civil or military.

Questions

1. In the middle of a 5 kHz channel adjacent to that carrying a wanted LM signal at a level of −90 dBm is RF pollution consisting of an unmodulated carrier at −30 dBm. The on-frequency local oscillator power is +24 dBm. What must be the limit on total local oscillator noise power in the adjacent channel to ensure at least 20 dB S/N ratio in the wanted channel. (−56 dBm)

2. A transmitting/receiving site has transmitters of 100 W power. ‘Rusty bolt’ effect produces third-order products with a conversion loss of 160 dB. What must be the minimum signal power to a receiver to guarantee a 30 dB signal/interference ratio (ignore feeder losses)? (−80 dBm)

3. A direct sequence spread-spectrum system transmits BPSK. When the S/N ratio at the receiver input is −6 dB the S/N ratio is +20 dB immediately prior to the modem. How many more transmitters of identical power and range can operate in the same band, on orthogonal codes, before the raw BER exceeds 10^{-4}? (42) If the data rate is 50 kb/s, what is the minimum bandwidth of the radio system? (20 MHz)

4. A military radio in a 30 kHz channel is attacked by a helicopter-borne noise-modulated jammer at a slant range of 3 km, which reduces its received S/N ratio to 0 dB. The radio is replaced by a frequency hopping equipment which hops a channel of the same width over a 3 MHz band. The jammer has its bandwidth widened to cover the hopping band as a countermeasure, but its total output power can only be doubled. How much nearer must the helicopter be to the target radio to achieve the same jamming effect, assuming that it flies high enough for free-space propagation? (slant range of 424 m)

CHAPTER 11

Taking the time

We have considered the influence of space and format, specifically sequency, on the design of radio systems. The other dimension available to the designer is **time**. Signals which arrive at the receiver at different times cannot interfere with each other, provided that the receiver has a good enough transient response and certain other minimal conditions are met. This can be exploited in its simplest form through transmission schedules, allocating different time slots through the day or year to different users, but more interestingly it can also operate on a much faster basis.

Time division has been widely used in radio systems, in this respect following line telecommunications practice. During the 1950s, when their adoption was first being taken seriously, time division systems were particularly appreciated for the simplicity of the hardware required to implement them, which was consistent with the relatively primitive state of technology at that time. This is no longer a dominant consideration, and their survival today is largely a matter of technological inertia.

11.1 Time division multiplexing

The earliest use of time displacement to achieve orthogonality in radio transmission paths was **time division multiplexing** (**TDM**). If several communication signals are to be carried simultaneously by a single radio channel, an obvious way to go about it is repeatedly to

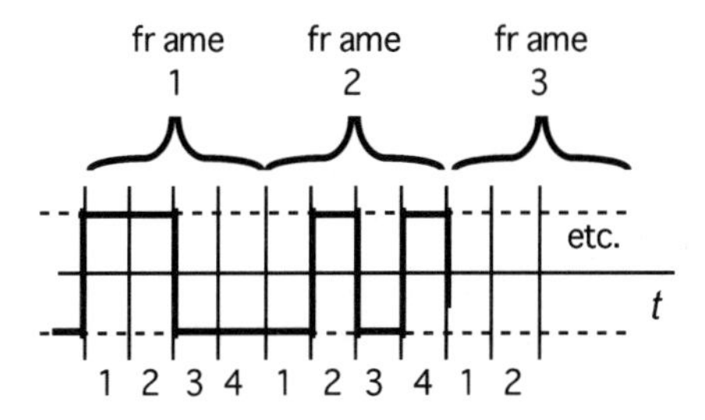

Fig. 11.1
Four-channel TDMA. In the two frames shown the signals transmitted are: channel 1: **1 0**, channel 2: **1 1**, channel 3: **0 0**, channel 4: **0 1**.

sample each signal in turn, always in the same order. Each message stream is broken into a series of short time segments (easy if the message is in digital form) and the segments from different message streams are interleaved in a regular sequence before transmission. The repeated sequences of segments may then be transmitted over a single radio channel. This process is known as **multiplexing (mux)**. Both analogue and digital signals may be transmitted in this way, although in the case of analogue signals it is usual to sample frequently enough for each sample to consist of just a single instantaneous value. Digital signals may be sampled one bit at a time or (much more rarely) each sample may be a short digital word. Figure 11.1 illustrates in simplified form the technique for the case of four signals multiplexed on to a single channel. Additional timing signals may be included in a practical system.

At the receiver the segments (or samples) are separated out and the individual message streams reassembled. This is **demultiplexing (demux)**. The receiver must demultiplex synchronously with the transmitter multiplexer, and it is common practice to transmit a **synchronizing pulse** (often distinguished either by greater amplitude or width) before each group of samples – the group along with the synchronizing pulse is sometimes called a **frame**.

In the special case where the same radio channel is exploited alternately for transmission in one direction and then in the reverse, to permit two-way communication using only the single channel, the technique is known as **time division duplexing (TDD)**.

A further development of TDM is **time division multiple access (TDMA)** used in cellular radiotelephone and other systems, whenever a central base station wishes to communicate simultaneously with many other (usually mobile or hand-portable) users.

A base station communicates with mobiles by assigning to each mobile calling in one of a number of time slots on an '**up**' or incoming (mobile-base) channel, and similarly another on a '**down**' or outgoing (base-mobile) channel. These time slots are held for as long as the connection continues, but at the termination of traffic they are reclaimed and reallocated to the next caller. Alternatively just a single channel may be used for both 'up' and 'down' traffic, with the different time slots assigned to 'up' and 'down', somewhat in the manner of TDD. Messages are time segmented, as before, and transmitted in sequence, just as in TDM. However, the mobile users receive only the signals in 'their' time slot, and similarly transmit back to the base station in only one of the cyclically repeated time slots, their 'own' assigned slot.

11.2 TDM and FDM compared

At one time TDM was seen as having a marked advantage over FDM (which does exactly the same job, needless to say). Up to about a quarter of a century ago the frequency-domain filters required for FDM systems had to be constructed using either combinations of wound inductors and capacitors, or else electromechanical resonant components, such as quartz crystals. Either approach was bulky, heavy, inflexible and very expensive compared with the simple hardware required for TDM, which consisted of little more than transistor gating circuits, along with integrators in the analogue case. As a consequence, and in order to minimize hardware costs, TDM entirely replaced FDM in-line telecommunications service and it has also had powerful advocates in radio use, being adopted for both the GSM mobile phone system and the DECT cordless telephone standard. However, technology moves on, and today digital signal processing has dramatically reduced the cost of implementing FDM (and its close relative CDM). How do the TDM and FDM systems compare, then, in a twenty-first century radio-use environment?

The most important difference between the line telecommunications and radio environments is that in the latter multipath propagation is very common, in fact almost universal. Provided that reflections are properly suppressed, in an unswitched line or cable system the propagation time between a transmitter and receiver has a unique stable value. By contrast, radio transmission is commonly simultaneous over more than one path. The examples of this are numerous, and have already been described. In all bands from MF upward, and particularly in the VHF/UHF scattering environment, multipath propagation is a very widespread phenomenon. Where it occurs it cannot be assumed that the first-arriving signal is much stronger than later signals, because distant reflectors (such as hills and large buildings) may be much bigger than close-in ones, and thus reflect more energy even although it is longer delayed. To see the effect of multipath on TDM systems, consider a digital signal which arrives over two paths (1) and (2) with propagation times T_1 and T_2. Suppose that the transmission is digital with a rate of B bauds (symbols per second). Clearly if:

$$B \approx \frac{1}{T_2 - T_1} \qquad (11.1)$$

the delay between the two paths is of the order of the intersymbol interval, and two different signals will arrive at the receiver at the same time, so an error is highly likely. To ensure that errors do not occur from this cause it is necessary that the delay should be much shorter than an intersymbol period so:

$$T_2 - T_1 \leq \frac{k}{B} \quad \text{where } 0 < k \ll 1 \text{ and typically } k \approx 0.1 - 0.2$$

or we may write, in terms of path lengths rather than time delays:

$$B \leq \frac{k}{T_2 - T_1} = \frac{k \cdot c}{L_2 - L_1} = \frac{k \cdot c}{\Delta L} \qquad (11.2)$$

The possible difference in path lengths depends on the use environment, ranging from as much as 100 km for HF sky-wave transmissions, through 1 km in the UHF mobile service down to as little as 10 m at EHF. The corresponding limiting symbol rates (taking $k = 0.2$) are respectively 600 bauds, 60 kbauds and 6 Mbauds. Multipath time stagger of received pulses can be improved, however, by using a **channel equalizer**. Suitable mathematical processing makes it possible to reduce the errors arising from the signal's interaction with a time-delayed version of itself. A number of algorithms have been proposed for this service, the most successful and widely used being the **Viterbi equalizer** which, although complex, can better than double the carrying capacity of a radio link. Nevertheless, in any propagation environment the signalling rate has a calculable limit, which can be extended but not evaded.

This has a bearing on the choice of multiplexing system. If N channels, each having a signalling rate of B_{CH} bauds, are transmitted together using TDM the combined signal will have a rate of at least NB_{CH} while if FDM is used N channels will be occupied each having a signalling rate of B_{CH} bauds. Usually the FDM system does not need channel equalizers but the TDM system does, in which case the perceived advantage of TDM, namely the simplicity of its hardware, is lost so far as the receiver is concerned. At the transmitter, modern practice is to generate both TDM and FDM waveforms at low level and then amplify them to the required power for transmission using linear amplifiers, so there is no advantage for either of the two systems of multiplexing in this respect.

Broadly speaking, there is not much to choose between FDM and TDM systems, although in environments where there is little or no multipath transmission TDM should be simpler, whilst FDM will perform better where mutipath problems are severe. Neither shows any particular advantage over the other in economy of spectrum use. TDM was more fashionable than FDM, largely because engineers with a line telecommunications background find it more familiar. Both are now increasingly eclipsed by code division mutliplexing (CDM) which gains considerable system advantages by generalizing frequency into the hyperdimensional sequency domain.

11.3 Asynchronous transmission

All of the time-sharing systems mentioned so far are fully **synchronous**, that is there must be clocks at both the receiving and transmitting ends of the links which give the same time to close tolerances, so that the correct **time slot** can be selected in each successive **frame** (cycle of slots). Often special signals are included in the transmissions to allow the receiver clock to be synchronized on start-up with that at the transmitter. In synchronous systems messages are received continuously and in real time, just as they are transmitted.

However, **asynchronous** time-sharing systems are also possible, and growing rapidly in importance, because they are far more flexible than the synchronous forms. The most important of these use **packet transmission**. A packet network, whether cabled or radio, acts like a main highway, capable of providing fast and reliable communication at low cost to multiple destinations. Although in principle it would be possible to conceive of packet analogue transmission, in practice all current and prospective packet systems transmit digital signals. The most widely used and familiar of all packet communication systems is the **Internet**, which is itself the successor to the **ARPA net** (Advanced Research Projects Agency) set up by the US Department of Defense, the first successful packet network of any size ever to be built. Its original purpose was to provide a means of communication which stood some chance of surviving nuclear warfare.

The essence of a packet system is that messages to be sent are broken up into short segments, known as **packets**, which are transmitted independently of each other. Obviously as well as the piece of information the packet contains, it must also carry other data in order to be useful, and at least the address to which it is to go and the time at which it was transmitted. Given this information the packets can be routed to the correct destination and assembled in the correct order, so that the segmented message can be reconstructed. In practice yet more information is almost invariably also incorporated, such as the sender's identity.

Clearly, for a packet system to work successfully all the packets must be created in a uniform standard format, which can originate at the transmitting location and be correctly interpreted when received. A number of protocols have been proposed for packet systems, each best suited to a particular range of applications. As an example of these, X.25 is a widely used standard for moderate speed data transmission. The make-up of the X.25 packet envisages a frame which begins with a **flag** (a start signal), followed by a **frame check sequence**, then the **user data** (128 bits), after that **packet control data**, a **control field**, the **address** and a final **flag** to signal the end.

Initially packet systems were used only for data transmission at modest rates, and this is still their dominant use in radio systems, but there is no reason in principle why packets should not be sent at a high enough rate for digital speech, or even video transmission. This is already happening in cable and satellite circuits and bids to become widespread. It is the basis of telephony on the Internet, already much used.

The great virtue of packet systems is their remarkable flexibility. Because all data carries the address to which it is directed it is not necessary to segregate it in a separate channel throughout its journey. A common data 'pool' can be used, into which packets can be injected from multiple sources and from which they can be retrieved by many recipients, identifying their own addresses. This principle, on which the Internet operates, can be a great deal cheaper than any other way of transmitting the same quantity of information, and avoids the need for circuit switching. Packet systems can also be made remarkably damage resistant, since it is permissible for packets to take alternative paths through complex networks, and the functionality can therefore survive the loss of some of the network nodes or links.

11.4 Radio traffic in the time domain

Aside from TDM and TDMA, which are means of preventing interference between simultaneous transmissions, the time domain plays an important role in communication because (aside from broadcasting) all messages are discontinuous time sequences of segments, either digital symbols or speech utterances. This gives an important opportunity for spectrum use economies by employing each radio channel to carry several, perhaps many, distinct message streams at different times.

Messages have specific durations, which are highly variable from one to another, and occur at time intervals which are often quite random. Thus typically communications channels are occupied at times and at other times are left free. A crucial factor in the design of systems is the **traffic density** (measured in **erlangs**). The traffic density is defined as:

$$R = \frac{\text{Mean message duration}}{\text{Average time between message starts}} = \frac{\delta}{\zeta} \qquad (11.3)$$

In a single channel the traffic density cannot quite reach unity, but we may also measure total traffic density over a bundle of channels, in which case the traffic density may be much greater than one erlang.

Obviously since each channel in a radio system occupies some electromagnetic spectrum, maximum economy of spectrum use is attained if every channel has a traffic density as near as possible to one erlang. Many simple two-way radio systems are very far from this ideal. For example, tests conducted around 1960 in central London showed that channels then used for public emergency services (police, fire, ambulance) had typical traffic densities of only 0.04 erlang. These studies led to the search for more efficient use of channel capacity and hence to **trunked** radio systems, now adopted very widely by this type of user.

In the simplest two-way radio system users are permanently assigned a channel each, which they use at will to pass traffic. Most of the time they will not be communicating, which leads to very low traffic densities. In a trunked system the number of channels used is less (and perhaps much less) than the number of users. When users wish to pass traffic they are temporarily assigned a channel for their exclusive use (in the usual case of frequency division duplexing, one each for up- and downlinks) for the duration of that contact only. Thus all channels carry traffic for many users, and can be far more heavily loaded. The economy in spectrum use made possible is large; however, even aside from a modest increase in hardware complexity, there is a price to be paid. With a trunking system there is no certainty that when a user wishes to initiate communication there will be necessarily be a channel available, so there is a small risk of being unable to communicate. In this case some arrangement is usually made for the user to receive a 'busy line, try later' signal, visual or audible.

Basically the trunked system has additional 'hailing' or signalling channels (one 'up', one 'down'), used only for system control. A mobile user wishing to pass messages contacts the base station via the hailing channel and is assigned 'up' and 'down' channels, either sequency assignments (FDMA, CDMA) or time slots (TDMA) – a fully automatic procedure. The base station can do this because it keeps a running register of the channels in use. The mobile then moves to the assigned channel pair and passes its traffic, retaining these channels for the duration of that contact only and relinquishing them when traffic ceases, either by signalling completion to the base station (which is invariably automatic) or by a time-out to allow for equipment failure (traffic ceases for a specified interval). In the event that a mobile wishes to initiate traffic but all channels are in use at that time the base station sends an engaged signal via the hailing channel. For base-to-mobile contacts the procedure is the same, except that the base station can instruct the mobile on which channels to use during its initial contact.

Trunking results in much improved channel loading. Suppose that there are N mobile users working to a base station with n working channel pairs, that is $(n + 1)$ including hailing channels, and each mobile originates r erlangs of traffic. We can speak of the possible states of the base station as being from none to n channels occupied and carrying traffic, the latter state being the fully loaded condition. So if s channels are occupied, then $0 \leq s \leq n$. A statistical model of the communications traffic problems assumes:

- that the amount of traffic offered by the users is independent of the number of radio channels currently in use, and

- that the system is at equilibrium, that is that the average traffic carried by the channels is constant.

Now if $p(s)$ is the probability of being in state s, then for equilibrium the rate of increase in number of channels used (due to new connections) must be balanced by the rate of decrease (due to connections ending), so if the number of users seeking to pass new traffic is h, and the number terminating traffic is k, then:

$$h(s-1) \cdot p(s-1) = k(s) \cdot p(s) \tag{11.4}$$

However, we have assumed that the mean call arrival rate is not dependent on the state of the base station, so:

$$\left[h(s) = h = \frac{Nr}{\delta}\right]_{\text{all } s} \tag{11.5}$$

also it is obvious that:

$$k(s) = \frac{s}{\delta} \tag{11.6}$$

so from equation (11.4):

$$\frac{s}{\delta} \cdot p(s) = \frac{Nr}{\delta} \cdot p(s-1)$$

and writing $Nr = R$:

$$s \cdot p(s) = R \cdot p(s-1) \tag{11.7}$$

Putting in values:

$$p(1) = \frac{R}{1} p(0), \quad p(2) = \frac{R^2}{1.2} p(0), \quad \text{and so on.}$$

In general:

$$p(s) = \frac{R^s}{s!} p(0)$$

However, the base station must be in one or other of its states, so the sum of the probabilities of all the states is equal to one. Hence:

$$\sum_{0}^{n} p(s) = \sum_{0}^{n} \frac{R^s}{s!} \cdot p(0) = 1$$

so, where u is an integer:

$$p(s) = \frac{\dfrac{R^s}{s!}}{\displaystyle\sum_{u=0}^{n} \frac{R^u}{u!}} \tag{11.8}$$

The base station will be unable to accept another request for a channel from a user, and hence traffic will be lost, when all n channels are in use. The probability of losing a call is therefore given by equation (11.8) when $s = n$, so the loss probability:

$$p_{\text{loss}} = \frac{\dfrac{(Nr)^n}{n!}}{\displaystyle\sum_{u=0}^{n} \frac{(Nr)^u}{u!}} \tag{11.9}$$

Factorial numbers rapidly become very large (factorial 10 is 3 628 800, for example), so this famous expression (the **Erlang loss formula**) is tedious to calculate for any but small n. Values of p_{loss} for the range of parameters interesting in radio system design can be plotted (Fig. 11.2). It is pointless to try computing values more closely than can be obtained from this graph, since they must be quite approximate because of the assumptions on which the formula is based. In particular it is not true that the rate at which users make calls is independent of the congestion of the system, but models which aim to take this into account realistically turn out too complex to be useful.

The effects of trunking are so dramatic that minor inaccuracies in estimating loss probabilities are rarely significant. Suppose, for example, that there are only two communicating channel pairs ('up' and 'down'). Without trunking they will accommodate two

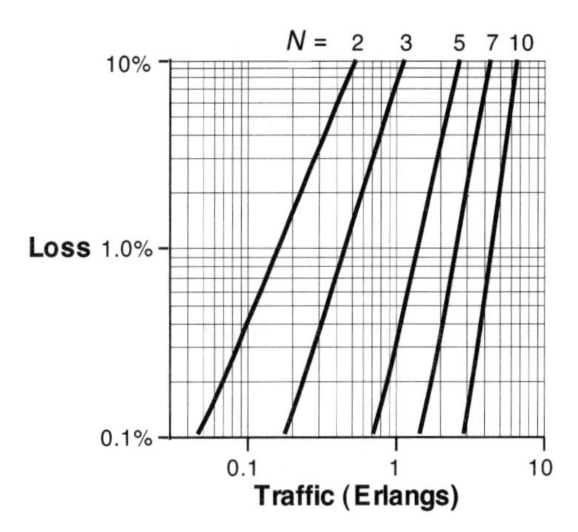

Fig. 11.2
The Erlang 'B' formula: probability that a channel will not be available versus traffic.

mobile users. If each of these originates just 0.05 erlangs – a typical figure – then the total traffic carried is 0.1 erlang. If these two channels are now used in a trunking system and a probability of 95% is set as the requirement for channel availability on a random call (one call in twenty fails), the total traffic carried is 0.36 erlangs before this limit is exceeded, so the channels will carry 0.18 erlang each, and the base station would accommodate seven mobile users passing 0.05 erlangs each. More than three times as many users are now operating satisfactorily, with only a 50% increase of spectrum occupancy (for the hailing channel). With more channels in use even greater advantages are obtained. In a 10 channel system the number of users, similarly calculated, goes up from 10 to 120 for a 10% increase in spectrum occupancy (assuming one hailing channel pair), the channels each carrying 0.6 erlang of traffic.

The on-line management of trunking systems needs little embodied computing power and only a modest software input, so the saving in spectrum occupancy is realized at minimal cost. This has led to widespread adoption of trunking as a component of many contemporary radio systems, including private mobile radio (PMR), cellular telephones and low Earth orbit satellite systems (LEOs), and often in combination with other spectrum conserving technologies. Only if the traffic originated from each user is unusually high, or the acceptable rate of call failure probability is very low, does the advantage of trunking become insufficient to justify its use.

11.5 Time conflict and its resolution

Efficient spectrum utilization requires that as many users as possible occupy as few channels as possible; trunking is one way in which this objective is achieved. When messages are very short or infrequent it may be decided that many users may access a single channel, for example the hailing channel in a trunked system. Much the same is true of the RF interface in paging systems. When many users try to access a system through a single radio channel like this, traffic will not be passed satisfactorily in event of time-based conflict between users. This will occur either when multiple users try to access the channel at the same time or when one user tries to access the channel while another is already using it. The latter circumstance is preventable. For example, in a typical mobile situation the base station may radiate a permissive signal in the 'down' (base-to-mobile) channel when it is not engaged by a mobile user, so if a user wishes to contact the base it does so only when the permissive signal is received. This system is known as **CSMA** (carrier sensitive multiple access), since the permissive signal is often simply the absence of base station carrier.

However, simultaneous access clashes cannot be avoided in this way, and a strategy has to be devised to deal with this situation. Contention for access between users will result in one or both failing to contact the base, which the user will confirm by failure to receive a response. Evidently, another attempt to contact the base station must be made, but when? In one commonly adopted strategy, the unsuccessful user waits for a time interval of randomized duration before making the next attempt; this is known as the **ALOHA** protocol. It works well provided that the traffic on the called channel is light, but a rapid increase of call blockage and repeats develops when the traffic exceeds 0.3 erlang per channel.

Attempts to access the base station by strong signals from a nearby user while weak signals from a distant user are being received can cause malfunction by interference or blocking the receiver. In an alternative version, called **slotted ALOHA** and particularly useful when (as in many digital systems) message durations are fixed (e.g. system management information packages of fixed format), the random time interval before recall is quantized rather than continuous. Thus the delay is $n \cdot \tau$ where n is a random integer and τ is a fixed time interval equal to the maximum expected message duration. Usually the base station signals the start of each time interval. The advantage of slotted ALOHA is that no access attempts are made during the passing of a message. Here too the utility of the system is confined to moderate traffic densities, although it can almost double the utilization of simple ALOHA.

11.6 Store-and-forward and burst transmission

In store-and-forward transmission, as the name suggests, received messages are stored – easy if they are in digital form and particularly so if they are packetized. They are then retransmitted when appropriate channel capacity is available. An early example was the use of this technique with low Earth orbit satellites (LEOs), which could receive transmissions from a ground station on one part of the Earth's surface, store them, and retransmit them back to ground at a later time when they were in a quite different part of their orbit, so that the recipient might well be a continent away from the point of origin of the message.

Another important use of store-and-forward is in meteor-scatter links, where the radio path is only occasionally and briefly available, so traffic must be stored until the link opens, when it is quickly transmitted at a high data rate. This is an example of **burst transmission**. A very short transmission (in the extreme case lasting as little as a few milliseconds) is made at very high data rates. In meteor-scatter communication the unpredictable nature of the channel makes this way of working unavoidable.

Burst transmission is also attractive to the military. If the burst is sent at a pre-arranged time, low errors in transmission often being ensured by using high power, friendly receivers can be ready to receive it. By contrast, the enemy is likely to be taken unawares, and will therefore, be able not to jam, locate nor intercept the transmission. All military burst transmissions necessarily adopt a store-and-forward strategy.

All store-and-forward systems involve a time delay between first transmission of the message and its ultimate reception by the designated recipient. In the case of LEOs this is predictable, but in most other cases it is not. In those applications where such a delay can be tolerated store-and-forward makes it possible to pack data into each channel up to a traffic density very close to one erlang when active. It is thus very conducive to maximum economy in spectrum utilization, but of course in many cases the delay, which can be protracted and unpredictable, is seen as an unacceptable price to pay.

Questions

1. VLF signals may propagate in a low-loss waveguide mode over the Earth's surface, so at a distance they may arrive either directly or around the Earth in the opposite direction. Assuming an omnidirectional receiving antenna and taking k (above) as 0.2, what is the limit on data rate due to multipath at a range of 5000 km (take Earth's diameter as 6378 km)? (6 bauds)

2. What do you understand by packet transmission? Describe a typical packet as used in radio systems. If speech can be encoded in 9.6 kb/s, at what rate would packets with X.25 format need to be transmitted to sustain a speech channel? (75 packets/s)

3. A trunked radio system has 40 mobile users each producing 0.05 erlangs of traffic and has seven channels carrying traffic. Assuming no congestion on the control channel, what proportion of attempted calls would you expect to be lost? (0.33%) If two of the radio channels fail what will be the new rate of loss? (3%)

Part Three

Spectrum Conserving Systems

CHAPTER 12

Spectrum conserving radio systems

We have considered a variety of technologies which make possible the use of the electromagnetic spectrum in a resource conserving way. In actual systems many of these are combined together, and to illustrate this we now consider three actual radio systems.

The first is **Digital Audio Broadcasting** (**DAB**) now replacing the outmoded FM broadcasting which is so remarkably wasteful of spectrum. Along with digital television, DAB will provide an effective solution to the problems of providing a high quality broadcasting service in a spectrum conserving manner. The second example is **packet radio (PR)** which grew out of a military requirement but soon attracted wider interest. Finally we look at **cellular radio telephones**, which have created the most rapidly growing industry of all time and are in the process of changing telecommunications radically. We consider both the presently dominant **GSM** system and **IMT-2000** which will replace it.

12.1 The slow death of FM broadcasting: DAB

FM sound broadcasting is one of the most wasteful uses of the spectrum at present tolerated. Although the highest audio frequencies transmitted are below 15 kHz, nevertheless because of the deviation ratio of five requires a 250 kHz channel (allowing for guard-bands). Frequency assignment problems can mean that in the UK twenty-two different assignments are required to establish a single nationwide channel, which even then gives an S/N ratio and stereo separation both markedly inferior to the cheapest CD recordings. To remedy this unacceptable situation, broadcasting authorities have for some years sought a digital system for CD-quality audio transmission which will have the appropriate qualities to supersede FM.

In Europe the outcome is **Eureka 147 DAB**, designed and developed by a consortium of universities, research institutes, equipment manufacturers and broadcasting authorities, under the auspices of the European Union's **Eureka** technical support programme. Development was essentially complete by 1993, when the system was successfully demonstrated in several European countries. The BBC's DAB services began in London in September 1995, operating from five transmitters in a single frequency network, and were rapidly extended. The official launch of DAB took place in Berlin during September 1997.

Soon after the turn of the century DAB will function alongside FM throughout Europe. Market penetration by DAB may well prove similar to the growth in CD use, and if so 40% of the population should be using DAB by 2007, although for a long period the two systems must co-exist, because of the need to safeguard the large investment by the public in FM receivers. Only when these have become obsolete will FM broadcasting be discontinued. Eureka DAB, which will be enjoyed in most of the rest of the world as well, offers largely interference-free reception and improved coverage as well as text, picture and multimedia transmission in addition to conventional sound broadcasting.

Most importantly, it will make far more efficient use of the spectrum, so that many more programmes can be provided without increasing spectrum occupancy. A further advantage is that transmitter power levels are much less than for earlier systems of broadcasting, so the chance of spectrum pollution is greatly reduced.

12.2 Technical aspects of DAB

The original brief was for an all-digital terrestrial sound broadcasting system, capable of transmitting up to five nationwide high quality channels using the upper VHF, UHF or possibly low SHF bands. In view of the wide availability and public acceptance of CDs as a source of recorded music, it was believed that a similar subjective audio quality must be provided by DAB in all radio environments, including scattering propagation.

Obviously the design would be dominated by the data rate required for each audio channel. The CD (for which the design was laid down in the late 1970s) provides high quality digital audio at a data rate of 1.4 Mb/s for a single channel, but improvements in coding, using digital signal processing, have substantially reduced the rate which need now be used for the same subjective effect. High quality music channels (such as BBC Radio 3) with full stereo capability are encoded for DAB, by a compression technique known as Musicam, at 256 kb/s. A predominantly speech channel (like the BBC's Radio 4 and 5) can give subjectively comparable results with only 192 kb/s, and so-called 'popular' music stations (Radio 1 and 2) require 224 kb/s. If these transmissions were interleaved in a single channel, it would have a data rate exceeding 1.088 Mb/s.

However, a simple binary channel at this rate would have a symbol duration of less than 920 nanoseconds. The radio system is required to operate at VHF or above in a very 'cluttered' propagation environment, encountered by portable radios in city centres. The likely spread of arrival times from scattered transmissions will therefore greatly exceed this symbol duration (typically being several microseconds), and very high error rates must result. Such a system would be impractical. An alternative solution which might be considered would be to use multi-level, rather than binary, digital signalling. But if enough levels were used to provide the necessary long symbol duration (at least 1024 giving 10 bits per symbol and hence a symbol lasting for $10 \times 0.92 = 9.2$ microseconds), a very high S/N ratio would be required if the difference in levels was not to be obscured by noise. Even using multi-level quadrature modulation, the protection ratio required would be very high, in excess of 40 dB. This approach therefore cannot yield a

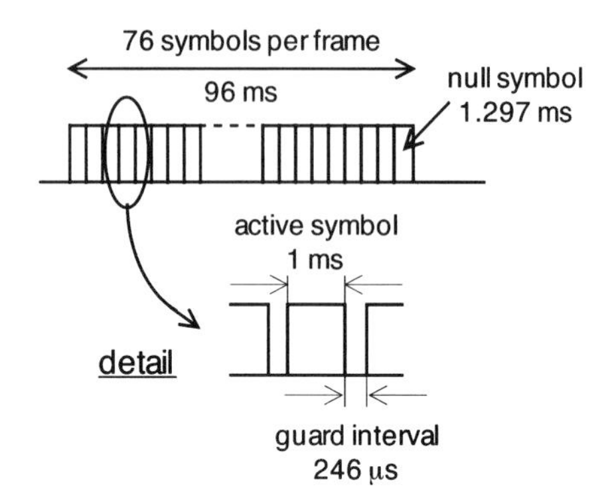

Fig. 12.1
DAB frame make-up.

satisfactory design either. At this stage in the evolution of the system it was clear that something quite new was required.

The solution finally adopted, known as **COFDM (Coded Orthogonal FDM)**, is to transmit a large number of parallel slow digital signals simultaneously, in FDM format on adjacent carrier frequencies, allowing the use of a very long symbol duration and extremely robust coding (Fig. 12.1). In all, 1536 independent carriers are used in adjacent narrow channels, with a carrier separation of only one kilohertz, so that a total of 1.536 MHz of spectrum is occupied. Each carrier has binary digital modulation in blocks of 96 milliseconds duration each, containing 76 active symbols and a null symbol to define block end. Each carrier transmits just under 792 symbols per second, and with 1536 carriers the total transmission rate is 1.216×10^6 symbols per second. The required minimum data rate is thus easily achieved by making each symbol equal to one bit, leading to robust binary signalling.

The 246 microsecond guard interval after each active symbol means that transmissions delayed by scattering will not have any effect on the next symbol, since a signal would have to travel an additional 80 km to exceed this time delay. Even if such an unlikely event were to occur as a substantial reflection from some object tens of kilometres distant, the signal arriving by this very long path would be too attenuated to be significant. Thus signal impairment due to multipath effects does not occur. In addition the binary modulation is so robust that adequately low transmission error rates are achieved at a channel S/N ratio of about 11 dB, so the protection ratio required is similar. With simulcasting, nationwide coverage for all five audio channels is given in the 1.536 MHz occupied; for comparison five FM broadcasting stations could require a total allotment of 20 MHz, to allow the assignment of different carrier frequencies in different geographical locations.

This use of a single assignment for DAB also has the advantage that no retuning of receivers is required when travelling in a car or taking a portable receiver from one area to another, since the same frequency is used at every transmitter. A single block of frequencies known as a **multiplex** carries six to eight DAB carrier groups and is radiated simultaneously, giving 30–40 distinct audio channels. In the radio spectrum 217.5 to 230.0 MHz, seven multiplexes have been allotted by the UK Government.

DAB technological development are already extremely sophisticated. Initially receivers were designed for the in-car market, but hi-fi and portable receivers are now available. PC card DAB receivers are also at an advanced stage, for reception of high-speed data, which may be encrypted for privacy. A tenth of all available DAB spectrum is assigned to data transmission, at a rate sufficient for feeding animated pictures to a colour monitor. The system, as well as giving better quality reception, has extreme spectrum economy for a service of this type. It is, of course, made practicable only by digital signal processing in the receiver, which makes it possible to demodulate so large a number of FDM channels in parallel using a single DSP chip working at a viable clock speed.

See: *http://www.worlddab.org*

12.3 Packet radio: a spectrum conserving system of military origin

Under battlefield conditions there is an overriding need for soldiers to communicate with each other despite enemy attacks on communications using **electronic counter measures** (**ECM**). There are also two other important requirements. The first is the obvious one that communications should survive even under conditions of considerable physical damage, which tends to penalize systems which are very hierarchical, with all-important nodes which the enemy might treat as priority targets. The second is that, as in all military communications, it is desirable to counter the enemy's **electronic surveillance measures** (**ESM**) by minimizing the transmitter power radiated. This reduces the chance of giving away valuable intelligence, whether by electronic signature identification (**elint**) or by actual interception of messages (**sigint**). It also minimizes the chances of position location by radio **direction finding** (**DF**), with the possibility of weapons targeting.

The military tactical radio requirement is often for an **all-informed net**, in which any participant can communicate with any other. Such a net realized by conventional radio coverage would have to be so organized that all participants were within the service range of all others, leading to relatively large service areas and transmitted powers (Fig. 12.2). The same net could be realized by achieving

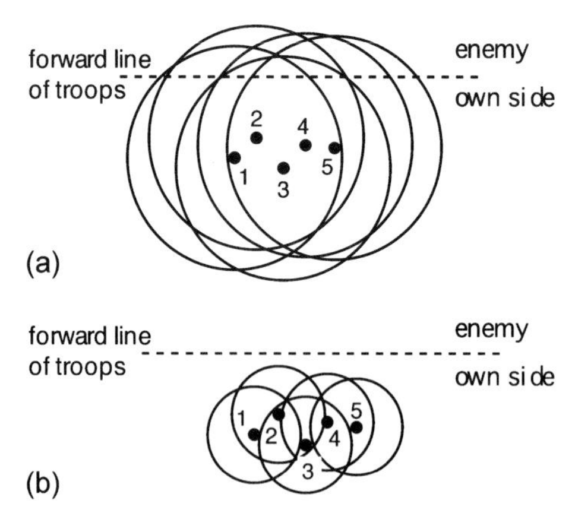

Fig. 12.2
A military all-informed net realized (a) conventionally and (b) using packet radio, where packets pass by a series of hops between adjacent stations. In (a) all stations can easily be intercepted or located by the enemy.

radio coverage in multiple hops, provided that each participant can communicate with at least one other and that the messages are packetized, so that they can be received and retransmitted by any member of the net. Since in practice the system will work at VHF or low UHF and will have a modest data transmission rate, this was at first taken to mean that speech could not be supported and the system would be data only. Although at one time speech communication was thought essential in all military tactical radio this is no longer the case, and battlefield data transmission is widely accepted. In any event, later developments have proved speech transmission to be practicable for packet radio.

With this type of multi-hop transmission the required transmitter range is much less, since the inverse fourth-power law relating transmitter output to range will normally apply, so the transmitter power can be greatly reduced. The interference and interception range is also reduced by the same factor so the system has **low probability of intercept** (**LPI**). The system is also **nodeless**, and can consequently be damage resistant if enough stations are present to ensure that each net participant can communicate with more than one other.

The packet design can be quite conventional: the header contains destination and other information, such as source and time of origin. The optional terminator is an end-marking code, used if the message is of variable length, and may also contain parity bits. An authentication code may be included in either header or terminator. Finally there will usually be a flag front and back. The data carried within the packet is invariably encrypted.

In typical packet radio hardware the normal state is the receive mode (Fig. 12.3). Received packets are held in a **FIFO** (first in, first out store) and retransmitted as soon as possible, in typical store-and-forward mode. Contention will inevitably arise in a fully populated system. This is dealt with conventionally (by ALOHA or slotted ALOHA, for example). With this simple packet radio system the packets would propagate to all stations within range. Thus the system could quickly fill up with traffic, including some packets propagating around loops. This can be overcome by using

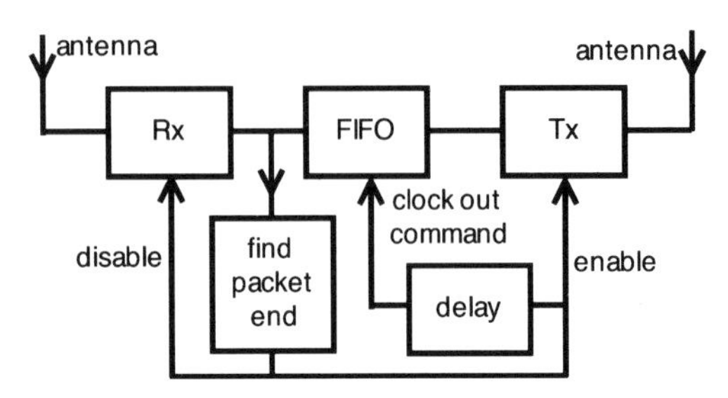

Fig. 12.3
Packet radio station configuration.

more intelligent stations which apply suitable protocols to packet handling, for example:

- Do not retransmit a packet originated by a station from which you have already transmitted a package with a later time of origin.
- If all the stations have knowledge of the positions of other stations (a system map), do not retransmit a package if you are vectorially further from the destination than the originating station.

Both of these protocols, though effective, will reduce the connectivity of the net. Indeed it is possible to think up pathological cases where legitimate messages will not get through intact if these rules are followed. There is always an unavoidable trade-off of loss of connectivity against improved system capacity through reduction of useless traffic.

The time delays and rate of transmission for a system of this kind are easily computed. If signals are transmitted at a data rate ρ_{raw} b/s and the packet is B bits long, of which b bits is message ($b < B$) it takes B/ρ_{raw} seconds to transmit and will take at least this long to read into the FIFO. A time t_{T} must then elapse for the receiver to turn off and the transmitter to power up, then the FIFO must be read out and finally the transmitter powers down again, which also takes t_{T}. Hence in a time τ given by:

$$\tau \geq 2\left(t_{\text{T}} + \frac{B}{\rho_{\text{raw}}}\right) \tag{12.1}$$

b bits of information are transmitted. The maximum possible transmission rate of the system is ρ, where:

$$\rho = \frac{b}{\tau} = \rho_{\text{raw}} \cdot \frac{b}{2(t_{\text{T}}\rho_{\text{raw}} + B)} \qquad (12.2)$$

Real systems will usually be worse than this because contention may further reduce the rate of transmission. The system is particularly inefficient if the transmitter change over time is not short compared with the packet length and if too large a proportion of the packet is taken up with non-message elements.

At first packet radio was used as a message system for data only. For example, if $\rho_{\text{raw}} = 16\,\text{kb/s}$, $B = 100$, $b = 60$ and $t_{\text{T}} = 1\,\text{ms}$ (which is fast), then $\rho = 4.14\,\text{kb/s}$, which was thought too low for speech transmission. However, low-rate voice codecs are now available with transmission rates below 2.4 kb/s, so simplex voice transmission by packet radio is practicable, and of considerable interest to the military user.

Subsequent to the military interest in packet radio, civil systems appeared. They are broadly similar, but the design responds to slightly different requirements. The military need for maximum damage resistance becomes less important in a civil system, so it is not essential for all users to be identically equipped, and some degree of hierarchy is permissible. The network therefore consists of terminal units with attached packet radios, repeaters (router infrastructure nodes), and stations (providing centralized administration). This makes it possible to keep the cost of terminal units lower than in the fully 'egalitarian' military version in which equipment comes in one form only, capable of performing all three functions (Kahn, Gronemeyer, Burchfield and Kunzelman 1978).

What has been described in this section is a true packet radio system, with multiple hops. Sometimes packet switching is used on other kinds of radio system, such as cellular radiotelephone, and is (unwisely) called packet radio. This is not the same thing at all, and should in this instance be called packet-switched cellular radio.

Packet radio is a nodeless system which greatly reduces interference and spectrum pollution (by reducing transmitter power) for an all-informed group of users. It can be made resistant to physical damage and ECM provided that there is sufficient propagation path redundancy. The reduced transmitter power results in low probability of intercept (LPI) as well as reducing the battery drain in portable equipments. All of these factors make the system attractive to military and public service users.

12.4 The most significant new system organization of all – cellular radio

Cellular radio is a radiotelephone system that appeared around 1970 and developed rapidly from the early 1980s. The area in which service is required is divided up into smaller zones, called **cells**, that have a radius of from a few hundreds of metres to a few kilometres. Each cell has its own transmitter (in some cases more than one, but operating synchronously). The 'honeycomb' pattern of cells can exploit frequency reuse to the maximum without interference, provided that an appropriate assignment pattern is adopted. Users equipped with radiotelephones can access the system or be accessed anywhere within the area of the ensemble of cells. When users move from one cell to the next, the transfer to the new channel frequencies, known as the **hand-off**, takes place with only a brief interruption.

Conventional radio systems used fixed frequency assignment, and the system, operating through a single transmitter, exchanges traffic only with the mobiles within its own coverage area, permanently losing contact with them if they move outside it. In a cellular system, by contrast, traffic for users anywhere in the whole area covered by the tiling pattern is transmitted via the fixed station(s) in the cell in which they happen to be situated at that time. For example, moving along the path u–v as shown, the mobile will receive signals from transmitters on the (outgoing) channels A, B, C and A in turn, according to the cell it is currently occupying (Fig. 12.4).

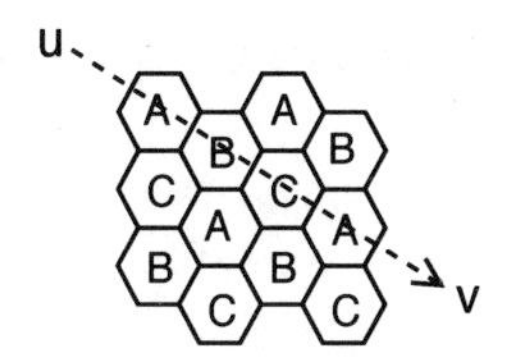

Fig. 12.4
In a cellular system a mobile travelling through transmits and receives on frequencies appropriate to the cell where it is currently located.

Similarly it will send back return traffic to fixed receivers in each of the cells on frequencies A, B, C and A (single frequency working) or A′, B′, C′ and A′ (double frequency working).

Thus a cellular system must 'know' the location of each mobile operating within it so as to be able to route outgoing traffic from the correct transmitter. This includes detecting when a mobile crosses a cell boundary so that transmission hand-off from one transmitter to another can be undertaken. This requires much more network intelligence than for any of the systems considered so far. Base stations are controlled by **mobile switching centres** (MSC) which also connect them to the public switched telephone network (PSTN) often via a centre which controls an area, where the system register is kept (Fig. 12.5). Both switching and control are computer intensive.

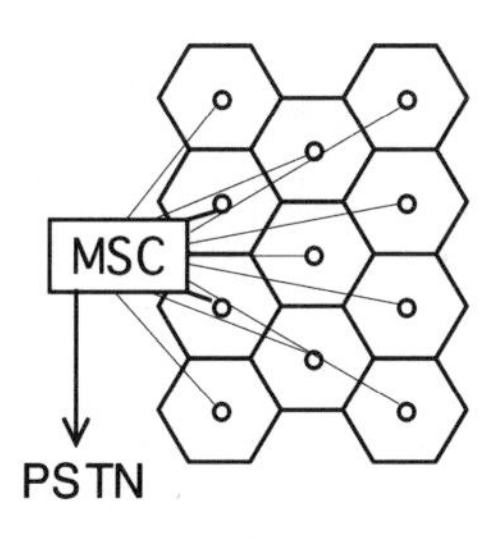

Fig. 12.5
The mobile switching centre (MSC) connects a group of cells to the public switched telephone network (PSTN), though maybe through an intermediate 'area' switch.

12.5 Cellular radio: the basic arithmetic

In designing a cellular system one of the most critical parameters is user density. If the probability of finding a user is p_{user} per square metre and the area of a cell is A, the expected number of users per cell is $p_{\text{user}}.A$. If the number of channels available is m, the mathematics then approximates to that for trunking with $p_{\text{user}} \cdot A$ replacing N in the expressions for p_{loss}, the probability of failing to make a call, in the expressions worked out for trunking and given in equation (11.9), namely:

$$p_{\text{loss}} = \frac{\dfrac{(p_{\text{user}} \cdot Ar)^n}{n!}}{\displaystyle\sum_{u=0}^{n} \dfrac{(p_{\text{user}} \cdot Ar)^u}{u!}} \tag{12.3}$$

where u is an integer, n the number of channels available, and r the traffic per user (in erlangs). The design of the system depends on choosing a value for n with an average sized cell at the expected user density to give the required (low) probability of being unable to place a call due to congestion. Cell sizes are then chosen so as to give, as far as possible, a constant value of $N = p_{\text{user}} \cdot A$, that is smaller A where user density is high (urban areas) and larger where the density is lower (rural). Base station transmitter power is adjusted to allow for the size of the cell served. (In practice even in relatively depopulated areas there is a limit to how large cells can be, which is set by maximum transmitter power or by terrain features.) Since the cell size is chosen to keep the expected cell occupancy constant, the expected revenues from service use in each cell are also roughly the same, so the same capital expenditure can be justified for any cell, regardless of size.

Note also that in principle any user density can be accommodated, without upper limit, by making cells small enough. At first sight this looks like a total solution to the problem of spectrum congestion, but in practice there is a lower limit to cell size, set by problems of hand-off. Consider a moving mobile passing through a series of contiguous cells with a velocity v. If the distance travelled within each cell is Λ the time spent in each cell is Λ/v. At each cell boundary hand-off will occur, which takes a time t' during which no signals are received (in either direction). There is therefore an error rate ε introduced by hand-off given by:

$$\varepsilon = \frac{v \cdot t'}{\Lambda + v \cdot t'} \tag{12.4}$$

For example, if the cell path length $= 10\,\text{km}$, velocity $= 30\,\text{m/s}$, hand-off time $= 1$ second (as in GSM), then the error rate is at least 3×10^{-3}.

The value of Λ depends on the mobile's trajectory, but is a monotone increasing function of cell size. Thus cell size cannot be indefinitely reduced without an unacceptable increase in raw error rate. For the purposes of system design, usually the worst case is considered, in which the path length in the cell equals the side–side dimension.

If the mobile velocity is low, for example portables carried by pedestrians, the floor on cell size is lower. Thus in the above example, for the same error rate with pedestrians walking at 1.5 m/s the path length in each cell could be 500 m. Even for road vehicles, when close together they are obliged to travel more slowly, which helps. However, problems arise when vehicle speeds in a given location vary widely at different times of day. Cell size chosen for low traffic density will probably suffer severe call loss at times of high density, as in the rush hour.

Another problem which sets a lower limit to cell size is **forced termination**. When a mobile crosses a cell boundary hand-off occurs, that is commands are received from the base station to switch to a different channel (or sequency for CDMA). However, in any real, populated system there is a non-zero probability that no channel will be available in the cell entered, all being in use. In this case there is forced termination, and communication is lost without warning. It is universally agreed that users regard this as a much worse defect than failure to access the system.

How large is the problem? The rate of crossing cell boundaries is v/Λ and at each such crossing there is a probability that forced termination will occur, say p_t which can be obtained from the Erlang traffic formula. The expected call duration is δ so the probability of forced termination during an average call is:

$$p_t = p_t \cdot \frac{\delta \cdot v}{\Lambda} \tag{12.5}$$

A forced termination probability during a call of as little as 1% is usually considered unacceptably high, which again sets a lower limit to cell size.

Because it is essential to route traffic to a mobile through the transmitter(s) covering the precise cell in which it is located, a call cannot be made to a mobile unless its location is known, so such information is stored by the system in a register (although mobiles can initiate calls to base at will without being registered). Since a cell phone may have moved, broadcast is performed over some number of cells around the phone's last known location.

The basis of the register is the location at the most recent time the mobile passed traffic, but since mobiles can be switched off for long periods during which they may move extensively, this is not enough alone, and search procedures are necessary, usually fanning out from the last known location. Various search algorithms have been proposed, mostly aimed at minimizing search time by using local storage of data and searching in parallel. As the number of mobiles, channels and cells increases the search time gets longer, and this is only partly offset by increasing search rate. An attractive solution is routine 'log-in' calls from mobiles on start-up and at regular times thereafter, of which the user need not be aware, and this is now general.

Some see the cellular system's unavoidable registration of user location as an invasion of civil liberties; access to the information is strictly controlled in most countries. It can have advantages – soon after cellular services began in Sicily a group of kidnappers were arrested in consequence of making their ransom demand by cellular phone, which revealed their position. However, it is possible that the ability to locate users might have a less benign outcome.

12.5 The evolution of GSM

Early cellular systems combine digital signalling with multiplexed analogue FM voice channels. Among these were NTS (Scandinavia), AMPS (US) and TACS, a closely related UK system. These systems were wasteful of spectrum as well as giving poor voice quality under marginal conditions. When it was decided that there must be an all-Europe system, in line with the development of European economic and political institutions, the opportunity was taken to adopt a more modern all-digital approach.

The pan-European Digital Cellular Radio System specified by the Groupe Spéciale Mobile, under the **European Telecommunications Standards Institute (ETSI)**, is now in the widest use. The system concept originated within CEPT as long ago as 1982 and the basic requirements had been established by 1985. Broad technical details of the radio access were settled by early 1987. Service introduction began in 1993. Originally **GSM** signified the initial letters of the specifying group, but it is now interpreted as **G**lobal **S**ystem for **M**obile telecommunications.

The GSM radio interface is a multicarrier TDMA scheme supporting eight channels per carrier in a macrocellular environment. Frequency division duplexing (FDD or two frequency duplex) is employed. A **RELP** (residual excited linear prediction) speech coding scheme is used to achieve low speech data rate. Convolutional error control coding protects encoded speech and control data. Interleaving of the digitized and encoded information over several coding time slots is employed to reduce the effects of error bursts over the radio channel. Slow frequency hopping is also incorporated to mitigate frequency dependent channel impairment. An important feature in GSM is the provision of equalization of the radio channel to overcome the detrimental effects of multipath fading. All this is summarized in Table 12.1.

GSM can support cell sizes ranging from over 35 km radius down to $\sim$ 1 km or less in dense urban scenarios, and uses the standard hexagonal grid reuse pattern. Hand-off in the GSM system is controlled by the **mobile switching centre (MSC)**. Signal strength measurements and cell congestion information are used by the

Table 12.1 GSM specification

RF carrier frequency	890–915 MHZ up-link 935–960 MHz down-link
RF carrier spacing	200 kHz
RF transmission rate	270.833 kb/s
Speech rate	13 kb/s RPE-LTP (RELP) Error protected up to 22.8 kb/s
Control channels	Several different types using separate time slots at different levels of packet build
Duplexing	FDD
TDMA	8 full rate speech time slots per frame
TDMA frame period	4.62 μs
Mobile peak output power	1–20 W
Mobile mean output power	125 mW–2.5 W

MSC to determine when a hand-off should occur, with the total hand-off process taking the of order ~ 1 s or more. GSM carries modified **ISDN** (**integrated subscriber digital network**) services (such as data, fax and so on).

In January 1989 the UK Department of Trade and Industry proposed a new service just below 2 GHz. This activity within the UK coincided with a strategic review of the mobile communications field by the Mobile Experts Group within ETSI. The outcome of these two activities was the decision to undertake standardization within the GSM activity of ETSI, under the name **DCS1800** (Digital Cellular System 1800 MHz). DCS1800 technical specifications are very close to the GSM specifications, although with minor differences.

In the USA cellular service was for a while limited to individual cities, but the industry has adopted a digital standard for countrywide services. There is limited use of GSM in the USA on the West Coast, but the pan-US standard did not turn out to be GSM, since a suitable block of spectrum was said not to be available countrywide. The cellular phone has had to be interleaved with existing services and is mostly relatively narrowband. This has slowed its development in the USA, where the market has not grown comparably with the GSM world.

See: *http://www.gsmworld.com*

12.6 After GSM: IMT-2000

Although GSM proved phenomenally successful world-wide, was adopted by over 100 countries and became a *de facto* standard everywhere outside the USA and closely associated territories (and even saw limited use there) it was inevitable that it must, in time, give place to a yet more advanced **third generation** system. This was the subject of specification by the ITU, and the new family of standards which defines the mobile telephone system of the early part of the twenty-first century is known as **IMT-2000**. From the beginning it was intended as a system to be adopted worldwide. Although there might be differences of system detail between geographical locations the design of the user terminals was to be such as to hide these from the user.

As the European contribution to the evolution of these standards a system was under development throughout the 1990s and into the twenty-first century, under the auspices of ETSI, known as **UMTS (Universal Mobile Telephone System)**. The start date for a UMTS public service (aligned, of course, with IMT-2000) was set as 2003 with general adoption two or three years later, but it was recognized that UMTS and GSM would continue to co-exist for a considerable time, in view of the excellence of the service that GSM already provides to most of its users. The design of UMTS began with the requirement that it should lead to:

- Worldwide compatibility, so that calls can be made and received in any location on Earth with the same equipment. Access can be through a terrestrial network of base stations or via an LEO.
- The same system functional in the home, office, cellular and satellite use environments.

- Medium speed data capability (2 Mb/s to static terminals, 384 kb/s to pedestrians and 144 kb/s to vehicles).
- Small, inexpensive user terminals.

It was an early conviction that cellular networks must evolve towards lower-powered transmitters, capable of operating in very small cells – **microcells** or **picocells** – to sustain very high user densities. This must imply frequent hand-offs, and hence a very much shorter hand-off time is imperative. **Soft hand-off** (make before break) is potentially a complete solution to this problem. If the system has a consistent universal frequency (or sequency) reuse pattern, a receiver capable of receiving signals in the cell about to be entered (as well as the one currently occupied) can be used to make soft hand-offs possible. This has most commonly been seen in proposed CDMA systems, but the idea is actually independent of the mode of modulation.

As for system of modulation, a meticulous study of the many competing possibilities was carried out by ETSI. Although TDMA seemed an obvious choice at the time that GSM was finalized, by the time the decision had to be made for UMTS the signal processing capability of chips had developed to the point where CDMA looked a far better option, for all the reasons detailed in Chapter 10. Cell frequency reuse patterns (as in TDMA and FDMA) are no longer strictly necessary once the move is made into the sequency domain; CDMA has been described as having a 'universal one-cell frequency reuse pattern', since orthogonality of transmissions in adjacent cells can be obtained in the sequency domain alone. There may, however, be compelling arguments for using more than one frequency assignment in particular cases, for regulatory reasons, to secure different propagation characteristics (for example, as between satellite and terrestrial use) or to reduce necessary protection ratios. Where small microcells and picocells are used within a larger cellular structure (in order to give particularly intense local coverage) use of different frequencies may be essential.

The chosen W-CDMA system uses a chip rate of 4.096 Mb/s, allowing each such non-sinusoidal carrier to fit within a total 5 MHz bandwidth. Each operator deploys three or four carriers.

On the 'down' link, most transmissions are received by users at only 10 kb/s (adequate for speech of good quality) but they are able to 'capture' more than one channel when they wish to receive data at higher rates. On the 'up' link users can transmit at 2 Mb/s.

Along with this advanced air interface, UMTS has many other innovative features, which it would be inappropriate to detail here. Significantly, the circuit switching of GSM is replaced by packet switching, for greater flexibility and much superior economy. It is beginning to seem as though circuit switching will soon join those other technologies – FM and TDMA among them – that once played a pivotal role in radio engineering but are now inevitably receding into history.

See: *http://www.itu.int/imt*

12.7 Conclusions

The ethical, spectrum-conserving techniques described in the earlier parts of this book have been applied to the advanced systems described in the present chapter with conspicuous success. No less positive have been the results when they have been exploited to create other radio and television systems. We have every reason, therefore, to be guardedly optimistic about the medium-term outcome of the continuing radio spectrum crisis.

Made practicable by the dramatic improvements in electronic hardware and software, and particularly the relentless advance of microelectronics technology, it is clear that the technical means are now available to accommodate all the presently foreseeable demands on the limited radio spectrum made by a developing human society worldwide. Doing so successfully will continue to depend upon high standards of professionalism, integrity and ethical awareness on the part of radio engineers responsible for spectrum regulation and the design of new systems.

Further reading

Alsebrook K. and Parsons D. 'Mobile radio propagation in British cities in the VHF and UHF bands' *IEEE Trans. Veh. Tech*, **VT26**, 1977.

Gosling W. 'A simple mathematical model of co-channel and adjacent channel interference' *Radio and Electronic Engineer* **48**, 1978.

Gosling W. *Radio Antennas and Propagation* Newnes, 1998.

International Telecommunications Union

Subjective assessment of sound quality
ITU-R Recommendation BS562-3, ITU, Geneva, 1997 (a).

Subjective assessment of conventional television systems
ITU-R Recommendation BS1128-2, ITU, Geneva, 1997 (b).

Determination of radio frequency protection ratios for frequency modulated sound broadcasting
ITU-R Recommendation BS641, ITU, Geneva, 1997 (c).

Radio frequency protection ratios for AM vestigial sideband terrestrial television systems
ITU-R Recommendation BS655-4, ITU, Geneva, 1997 (d).

Protection ratios for LF, MF and HF broadcasting
ITU-R Recommendation BS562-3, ITU, Geneva, 1998.

Kahn R. E., Gronemeyer S. A., Burchfield J. and Kunzelman, R. C. 'Advances in packet radio technology' *Proceedings of the IEEE*, **66**, 11 November 1978.

Maral G. and Bousquet M. *Satellite Communication Systems* (3rd edition) Wiley, 1998.

Withers D. J. *Radio Spectrum Management* Peter Peregrinus, 1991.

Zverev A. I. *Handbook of Filter Synthesis* Wiley, 1967.

INDEX